Springer Series in Advanced Manufacturing

Series editor

Duc Truong Pham, University of Birmingham, Birmingham, UK

The **Springer Series in Advanced Manufacturing** includes advanced textbooks, research monographs, edited works and conference proceedings covering all major subjects in the field of advanced manufacturing.

The following is a non-exclusive list of subjects relevant to the series:

1. Manufacturing processes and operations (material processing; assembly; test and inspection; packaging and shipping).
2. Manufacturing product and process design (product design; product data management; product development; manufacturing system planning).
3. Enterprise management (product life cycle management; production planning and control; quality management).

Emphasis will be placed on novel material of topical interest (for example, books on nanomanufacturing) as well as new treatments of more traditional areas.

As advanced manufacturing usually involves extensive use of information and communication technology (ICT), books dealing with advanced ICT tools for advanced manufacturing are also of interest to the Series.

Springer and Professor Pham welcome book ideas from authors. Potential authors who wish to submit a book proposal should contact Anthony Doyle, Executive Editor, Springer, e-mail: anthony.doyle@springer.com.

More information about this series at http://www.springer.com/series/7113

Alp Ustundag · Emre Cevikcan

Industry 4.0: Managing The Digital Transformation

 Springer

Alp Ustundag
Istanbul Teknik Universitesi
Maçka, Istanbul
Turkey

Emre Cevikcan
Istanbul Teknik Universitesi
Maçka, Istanbul
Turkey

ISSN 1860-5168   ISSN 2196-1735   (electronic)
Springer Series in Advanced Manufacturing
ISBN 978-3-319-57869-9   ISBN 978-3-319-57870-5   (eBook)
https://doi.org/10.1007/978-3-319-57870-5

Library of Congress Control Number: 2017949145

Printed on acid-free paper

This Springer imprint is published by Springer Nature
The registered company is Springer International Publishing AG
The registered company address is: Gewerbestrasse 11, 6330 Cham, Switzerland

Preface

As a new industrial revolution, the term Industry 4.0 is one of the most popular topics among industry and academia in the world. Industry 4.0 plays a significant role in strategy to take the opportunities of digitalization of all stages of production and service systems. The fourth industrial revolution is realized by the combination of numerous physical and digital technologies such as artificial intelligence, cloud computing, adaptive robotics, augmented reality, additive manufacturing and Internet of Things (IoT). Regardless of the triggering technologies, the main purpose of industrial transformation is to increase the resource efficiency and productivity to increase the competitive power of the companies. The transformation era, which we are living in now, differs from the others in that it not only provides the change in main business processes but also reveals the concepts of smart and connected products by presenting service-driven business models.

In this context, this book is presented so as to provide a comprehensive guidance for Industry 4.0 applications. Therefore, this book not only introduces implementation aspects of Industry 4.0, but also proposes conceptual framework for Industry 4.0 with respect to its design principles. In addition, a maturity and readiness model is proposed so that the companies deciding to follow the path of digital transformation can evaluate themselves and overcome the problem of spotting the starting point. A technology roadmap is also presented to guide the managers of how to set the Industry 4.0 strategies, select the key technologies, determine the projects, construct the optimized project portfolio under risk and schedule the projects in planning horizon. Meanwhile, the reflections of digital transformation on engineering education and talent management are also discussed. Then, the book proceeds with key technological advances that form the pillars for Industry 4.0 and explores their potential technical and economic benefits via demonstrations with real-life applications.

We would like to thank all the authors for contributing to this book

- Sule Itir Satoglu, Istanbul Technical University
- Basar Oztaysi, Istanbul Technical University
- Sezi Cevik Onar, Istanbul Technical University

- Gokhan Ince, Istanbul Technical University
- Ihsan Kaya, Yildiz Technical University
- Erkan Isikli, Istanbul Technical University
- Gaye Karacay, Istanbul Technical University
- Burak Aydin, Silver Spring Networks
- Omer F. Beyca, Istanbul Technical University
- Mehmet Bulent Durmusoglu, Istanbul Technical University
- Seda Yanik, Istanbul Technical University
- Selcuk Cebi, Yildiz Technical University
- Gulsah Hancerliogullari, Istanbul Technical University
- Mehmet Serdar Kilinc, Oregon State University
- Mustafa Esengun, Istanbul Technical University
- Baris Bayram, Istanbul Technical University
- Ceren Oner, Istanbul Technical University
- Mahir Oner, Istanbul Technical University
- Beyzanur Cayir Ervural, Istanbul Technical University
- Bilal Ervural, Istanbul Technical University
- Peiman Alipour Sarvari, Istanbul Technical University
- Alperen Bal, Istanbul Technical University
- Aysenur Budak, Istanbul Technical University
- Cigdem Kadaifci, Istanbul Technical University
- Ibrahim Yazici, Istanbul Technical University
- Mahmut Sami Sivri, Istanbul Technical University
- Kartal Yagiz Akdil, Istanbul Technical University

We would also like to thank our colleague Ceren Salkin Oner for her support to prepare the final format of the book. And finally, we thank our families for their moral support and endless patience.

Istanbul   Alp Ustundag
2017   Emre Cevikcan

Contents

Authors and Contributors

About the Authors

Alp Ustundag is a full Professor at Industrial Engineering Department of Istanbul Technical University (ITU) and the head of RFID Research and Test Lab. He is also the coordinator of MSc. in Big Data and Business Analytics programme in ITU. He had been responsible for establishment of Technology Transfer and Commercialization Office of ITU as an advisor to the Rector. He worked in IT and finance industry from 2000 to 2004. He is also the General Manager of Navimod Business Intelligence Solutions (http://navimod.com/) located in ITU Technopark, which is a software company focusing on data analytics and business intelligence solutions. He has conducted a lot of research and consulting projects in RFID systems, logistics and supply chain management and data analytics for major Turkish companies. His current research interests include data analytics, supply chain and logistics management, industry 4.0, innovation and technology management. He has published many papers in international journals and presented various studies at national and international conferences.

Emre Cevikcan is currently an associate professor of Industrial Engineering Department in Istanbul Technical University. He received the B.S. degree in Industrial Engineering from Yıldız Technical University, the M.Sc. degree and Ph.D. degree in Industrial Engineering from Istanbul Technical University. He studied the scheduling of production systems for his Ph.D. dissertation. His research has so far focused on the design of production systems (assembly lines, production cells, etc.), lean production, scheduling. He has several research papers in International Journal of Production Research, Computers and Industrial Engineering, Assembly Automation, Expert Systems with Applications, International Journal of Information Technology & Decision Making. He is currently a reviewer in OMEGA, European Journal of Operational Research, International Journal of Production Research, Applied Soft Computing, Journal of Intelligent Manufacturing and Journal of Intelligent and Fuzzy Systems.

Contributors

Kartal Yagiz Akdil is a fresh Industrial Engineer and he is a business developer and R&D member in Migros Ticaret A.Ş. He is involved in many projects in the retail industry and led a specific project about gaming and e-sport. He is also the co-founder of Coinkolik (http://coinkolik.com) which is a Turkish news resource on bitcoin, blockchain and digital currencies. Previously, he co-founded FullSaaS, the web-based directory focused on SaaS and cloud applications. Kartal received his B.S. in Industrial Engineering from Istanbul Technical University. Kartal speaks fluent Turkish and English.

Burak Aydin has a Mechanical Engineering degree from Middle Eastern Technical University followed by an MBA degree. He started his professional career working as a consultant at Andersen Consulting/Accenture in Germany and Austria offices between 2001–2003. He worked for Siemens Business Systems as a Strategic Planning Manager between 2003–2006. He joined Intel Corporation Turkey by 2006 and lead as Managing Director between 2011–2016, established Intel Turkey R&D Center on May 2014, focusing on Internet of Things (IoTs) technologies. By 2017, Burak Aydin joined Silver Spring Networks as a Europe Middle East and Africa (EMEA) General Manager.

Alperen Bal received the B.E. degree in Mechanical Engineering from Namik Kemal University, Tekirdag, in 2010, and M.Sc. degree in Industrial Engineering from Istanbul Technical University, Istanbul, in 2013 respectively. Since 2013, he has been a Ph.D. candidate in Industrial Engineering in Istanbul Technical University. His current research interest includes lean production systems and logistics and supply chain management.

Baris Bayram is a Ph.D. candidate in the Faculty of Computer and Informatics Engineering at Istanbul Technical University. He received his B.Sc. degree from Izmir University of Economics, and his M.Sc. degree from Istanbul Technical University. His major research interest is robot perception.

Omer Faruk Beyca received the B.S. degree in industrial engineering from Fatih University, Istanbul, Turkey, in 2007, and the Ph.D. degree from the School of Industrial engineering and Management, Oklahoma State University, Stillwater, OK, USA. He is currently an Assistant Professor with the Department of Industrial Engineering, Istanbul Technical University, Istanbul, Turkey. Prior to that, he was a faculty member with the Department of Industrial Engineering, Fatih University, Istanbul, Turkey. His current research interests are modeling nonlinear dynamic systems and quality improvement in micro-machining and additive manufacturing.

Aysenur Budak graduated from Industrial Engineering Department of Sabanci University in 2010. She got M.Sc. degree from Istanbul Technical University (ITU) in 2013 and continued her doctoral studies at the Department of Industrial Engineering of ITU, and currently she is a Research Assistant at ITU.

Selcuk Cebi is currently an Associated Professor of Industrial Engineering at Yildiz Technical University. He received degree of Ph.D. from Industrial Engineering Program of Istanbul Technical University in 2010 and degree of M.Sc. from Mechanical Engineering Department of Karadeniz Technical University in 2004. His current research interests are decision support systems, multiple-criteria decision-making, human–computer interactions, and interface design.

Mehmet Bulent Durmusoglu is a full Professor of Industrial Engineering at Istanbul Technical University. He obtained his Ph.D. in Industrial Engineering from the same university. His research interests are design and implementation of cellular/lean manufacturing systems. He has also authored numerous technical articles in these areas.

Beyzanur Cayir Ervural is a Research Assistant and Ph.D. candidate at Istanbul Technical University, Department of Industrial Engineering. Her major areas of interest include energy planning, forecasting, sustainability, multi-objective/criteria decision-making and optimization.

Bilal Ervural is a Ph.D. candidate and a Research Assistant at Industrial Engineering Department of Istanbul Technical University. His research interests include group decision-making, multiple-criteria decision-making, fuzzy logic applications, supply chain management, mathematical modelling and heuristic methods.

Mustafa Esengun studied computer engineering at the Middle East Technical University (METU) in Northern Cyprus (Turkey) and completed his M.Sc. at Computer Engineering Department of Istanbul Technical University (ITU). He is currently a research assistant at the Computer Engineering Department of ITU since 2014. His main academic interests are user experience of augmented reality interfaces and industrial applications of augmented reality technology. He is currently doing his Ph.D. on integrating AR solutions with industrial operations.

Gulsah Hancerliogullari is an assistant professor of Industrial Engineering at Istanbul Technical University. She graduated with B.S. and M.S. in Industrial Engineering, and a Ph.D. in Engineering Management and Systems Engineering. Her current research interests are empirical research in operations management, application of optimization methods to transportation and healthcare problems, inventory management and statistical decision-making.

Gokhan Ince received the B.S. degree in Electrical Engineering from Istanbul Technical University, Turkey, in 2004, the M.S. degree in Information Engineering in 2007 from Darmstadt University of Technology, Germany and the Ph.D. degree in the Department of Mechanical and Environmental Informatics, Tokyo Institute of Technology, Japan in 2011. From 2006 to 2008, he was a researcher with Honda Research Institute Europe, Offenbach, Germany and from 2008 to 2012, he was with Honda Research Institute Japan, Co., Ltd., Saitama, Japan. Since 2012, he has been an Assistant Professor with the Computer Engineering Department, Istanbul Technical University. His current research interests include human–computer interaction, robotics, artificial intelligence and signal processing. He is a member of IEEE, RAS, ISAI and ISCA.

Erkan Isikli is currently Lecturer of Industrial Engineering at Istanbul Technical University (ITU). He earned his B.Sc. in Mathematics Engineering from ITU,

Turkey, in 2004, and his Ph.D. in Industrial and Systems Engineering from Wayne State University, USA, in 2012. His research mainly focuses on "Product Variety Management" and "Statistical Modeling". Along with his research activities, Dr. Isikli has taught courses on probability, statistics, stochastic processes, experimental design, quality control and customer relationship management.

Cigdem Kadaifci completed her Bachelor's and Master's degrees in Istanbul Technical University—Industrial Engineering Department. She has been working as a Research Assistant at the same department since 2010. She continues her Ph.D. in Industrial Engineering Programme and her research interests include futures research, multiple-criteria decision-making, statistical analysis and strategic management.

Ihsan Kaya received the B.S. and M.Sc. degrees in Industrial Engineering from Selçuk University. He also received Ph.D. degree from Istanbul Technical University on Industrial Engineering. Dr. Kaya is currently an Assistant Professor Dr. at Yıldız Technical University Department of Industrial Engineering. His main research areas are process capability analysis, quality management and control, statistical and multiple-criteria decision-making, and fuzzy sets applications.

Gaye Karacay is an Assistant Professor at the Industrial Engineering Department of Istanbul Technical University. Her Ph.D. in Management and Organization is from Bogazici University with a focus on Organizational Behaviour. Before her Ph.D. studies, Dr. Karacay had a professional work experience at public and private sector institutions in strategic management and public management areas. She has an MBA degree from London Business School (LBS). Her research interests include leadership, cross-cultural management, organizational culture, human resource management, talent management and corporate entrepreneurship. She has publications in international journals including Journal of World Business and Personnel Review. She has presented her studies at several international conferences.

Mehmet Serdar Kilinc is a postdoctoral researcher at Oregon State University. He formerly worked as a postdoctoral researcher at the Pennsylvania State University. He obtained his Ph.D. degree in industrial engineering at the University of Arkansas. He graduated with bachelor's and master's degrees from Istanbul Technical University, Turkey. His primary research interest is developing quantitative approaches to design and evaluate healthcare delivery and IT systems.

Ceren Oner received her B.S. degree in Industrial Engineering Department from Çukurova University in 2011. In 2011, she started to work as a Research Assistant in Istanbul Technical University and is currently a Ph.D. candidate in the same university. She writes and presents widely on issues of location-based systems, data mining and fuzzy logic.

Mahir Oner received his B.S. degree from Istanbul Technical University, Industrial Engineering Department. He had experience in private sector as business development engineer, method engineering and planning engineer. Currently, he is working as a research assistant in Istanbul Technical University and he is a Ph.D. candidate in the same university. His main research areas are real time tracking systems, RFID and Industry 4.0 applications.

Sezi Cevik Onar is an Associate Professor in the Industrial Engineering Department of Istanbul Technical University (ITU) Management Faculty. She earned her B.Sc. in Industrial Engineering and M.Sc. in Engineering Management, both from ITU. She completed her Ph.D. studies at ITU and visited Copenhagen Business School and Eindhoven Technical University during these studies. Her Ph.D. was on strategic options. Her research interests include strategic management and multiple criteria decision-making. She took part as a researcher in many privately and publicly funded projects such as intelligent system design, organization design, and human resource management system design. Her refereed articles have appeared in a variety of journals including Supply Chain Management: An International Journal, Computers & Industrial Engineering, Energy, and Expert Systems with Applications.

Basar Oztaysi is a full-time Associate Professor at Industrial Engineering Department of Istanbul Technical University (ITU). He teaches courses on data management, information systems management and business intelligence and decision support systems. His research interests include multiple criteria decision-making, data mining and intelligent systems.

Peiman Alipour Sarvari is a researcher at Industrial Engineering department of Istanbul Technical University. His current fields of interest include machine learning, virtual experiments, data analytics, supply chain management and logistics. He has plenty of book chapters and papers on maritime safety simulation, frequent pattern mining, artificial intelligence and mathematical inferences.

Sule Itir Satoglu is Associate Professor at Industrial Engineering Department of Istanbul Technical University (ITU). She earned her Mechanical Engineering bachelor's degree from Yildiz Technical University, in 2000. Later, she earned her Engineering Management M.Sc. degree in 2002, and Industrial Engineering Ph.D. degree in 2008, from Istanbul Technical University. Her research interests include lean production systems and logistics and supply chain management.

Mahmut Sami Sivri is currently a Ph.D. Candidate at Industrial Engineering Department in Istanbul Technical University. He also received the B.S. degree in Computer Engineering and the M.Sc. degree in Engineering Management from Istanbul Technical University. He worked in various companies and positions in the software industry since 2008. His current research interests include big data and applications, Industry 4.0, financial technologies, data analytics, supply chain and logistics optimization as well as software development and web applications.

Seda Yanik is an associate professor in Istanbul Technical University (ITU), Department of Industrial Engineering. She earned both her B.Sc. (1999) and Ph.D. (2011) degrees in Industrial Engineering from ITU. She also worked at multinational companies, such as SAP and adidas. Her research areas include logistics and supply chain, location modelling, decision-making and statistical quality control. She has published many papers in top-tier journals such as European Journal of Operations Research, Knowledge-Based Systems, and Network and Spatial Economics.

Ibrahim Yazici has been research assistant for 5 years. He is doing Ph.D in industrial engineering at Istanbul Technical University. He received B.Sc. degree in Industrial Engineering from Kocaeli University in 2011, M.Sc. degree from ITU in 2015. His interest areas are multiple-criteria decision-making, data mining applications, business analytics.

Part I
Understanding Industry 4.0

Chapter 1
A Conceptual Framework for Industry 4.0

Ceren Salkin, Mahir Oner, Alp Ustundag and Emre Cevikcan

Abstract Industrial Revolution emerged many improvements in manufacturing and service systems. Because of remarkable and rapid changes appeared in manufacturing and information technology, synergy aroused from the integration of the advancements in information technology, services and manufacturing were realized. These advancements conduced to the increasing productivity both in service systems and manufacturing environment. In recent years, manufacturing companies and service systems have been faced substantial challenges due to the necessity in the coordination and connection of disruptive concepts such as communication and networking (Industrial Internet), embedded systems (Cyber Physical Systems), adaptive robotics, cyber security, data analytics and artificial intelligence, and additive manufacturing. These advancements caused the extension of the developments in manufacturing and information technology, and these coordinated and communicative technologies are constituted to the term, Industry 4.0 which was first announced from German government as one of the key initiatives and highlights a new industrial revolution. As a result, Industry 4.0 indicates more productive systems; companies have been searching the right adaptation of this term. On the other hand, the achievement criteria and performance measurements of the transformation to Industry 4.0 are still uncertain. Additionally, a structured and systematic implementation roadmap is still not clear. Thus, in this study, the fundamental relevance between design principles and technologies is given and conceptual framework for Industry 4.0 is proposed concerning fundamentals of smart products and smart processes development.

C. Salkin (✉) · M. Oner · A. Ustundag · E. Cevikcan
Department of Industrial Engineering, Faculty of Management,
Istanbul Technical University, 34367 Macka, Istanbul, Turkey
e-mail: csalkin@itu.edu.tr

M. Oner
e-mail: mahironer@itu.edu.tr

A. Ustundag
e-mail: ustundaga@itu.edu.tr

E. Cevikcan
e-mail: cevikcan@itu.edu.tr

© Springer International Publishing Switzerland 2018
A. Ustundag and E. Cevikcan, *Industry 4.0: Managing The Digital Transformation*,
Springer Series in Advanced Manufacturing, https://doi.org/10.1007/978-3-319-57870-5_1

3

1.1 Introduction

Since first Industrial Revolution had aroused after steam engine, the following radical changes were appeared such as digital machines, automated manufacturing environment, and caused significant effects on productivity. The main reasons and triggers of the radical changes are individualization of demand, resource efficiency and short product development periods. Thus, enormous developments such as Web 2.0, Apps, Smartphones, laptops, 3D-printers appeared and this situation creates a big potential in the development of economies. Recently, in European Union, almost 17% of the GDP is explicated for by industry, which also effectuated approximately 32 million job opportunities (Qin et al. 2016). In contrast to this potential, today's companies are dealing with the challenges in rapid decision making for increasing productivity. One example could be given from the transformation process toward automated machines and services, which leads the coordination and connection of distributed complex systems. For this aim, more software-embedded systems are engaged in industrial products and systems, thereby, predictive methods should be constituted with intelligent algorithms in order to support electronic infrastructure (Lee et al. 2015).

In parallel to the necessity of coordination mechanism, synergy aroused from the integration of the advancements in information technology, services and manufacturing forms a new concept, Industry 4.0, was first declared by German government during Hannover Fair in 2011 as the beginning of the 4th industrial revolution. As explained in Bitkom, VDMA, ZVEI's report (2016), an increasing number of physical elements obtain receivers such as sensors and tags as a form of constructive technology and these elements have been connected after then the improvements seen in Internet of Things field. Additionally, electronic devices connection is conducted as a part of distributed systems to provide the accessibility of all related information in real time processing. On top of it, ability to derive the patterns from data at any time triggers more precise prediction of system behavior and provides autonomous control. All these circumstances influence the current business and manufacturing processes while new business models are being emerged. Hence, challengers for modern industrial enterprises are appeared as more complex value chains that require standardization of manufacturing and business processes and a closer relation between stakeholders.

The term, Industry 4.0 completely encounters to a wide range of concepts including increments in mechanization and automation, digitalization, networking and miniaturization (Lasi et al. 2014). Moreover, Industry 4.0 relies on the integration of dynamic value-creation networks with regard to the integration of the physical basic system and the software system with other branches and economic sectors, and also, with other industries and industry types. According to the concept of Industry 4.0, research and innovation, reference architecture, standardization and security of networked systems are the fundamentals for implementing Industry 4.0 infrastructure. This transformation can be possible by providing adequate substructures supported by sensors, machines, workplaces and information technology systems that are communicating with each other first in a single enterprise and certainly with other communicative systems. These types of systems referred as cyber physical systems and coordination between these systems are provided by Internet based protocols and standards.

As seen from the improvements in production and service management, Industry 4.0 focuses on the establishment of intelligent and communicative systems including machine-to-machine communication and human-machine interaction. Now and in the future, companies have to deal with the establishment of effective data flow management that is relied on the acquisition and assessment of data extracted from the intelligent and distributed systems interaction. The main idea of data acquisition and processing is the installation of self-control systems that enable taking the precautions before system operation suffered. Thus, companies have been searching the right adaptation of Industry 4.0.

In this respect, transformation to Industry 4.0 is based on eight foundational technology advances: adaptive robotics, data analytics and artificial intelligence (big data analytics), simulation, embedded systems, communication and networking such as Industrial Internet, cloud systems, additive manufacturing and virtualization technologies. These technologies should be supported with both basic technologies such as cyber security, sensors and actuators, RFID and RTLS technologies and mobile technologies and seven design principles named as real time data management, interoperability, virtualization, decentralization, agility, service orientation and integrated business processes (Wang and Wang 2016). These design principles and technologies enable practitioners to foresee the adaptation progress of Industry 4.0. On the other hand, a structured and systematic implementation roadmap for the transformation to Industry 4.0 is still uncertain. Thus, in this study, the fundamental relevance between design principles and supportive technologies is given and conceptual framework for Industry 4.0 is proposed concerning fundamental links between smart products and smart processes. First, supportive technologies are defined by giving specific implementation cases. In this respect, design principles are matched with the existing technologies. Besides that, a conceptual framework for a strategic roadmap of Industry 4.0 is presented, consisting of multi-layered and multi-functional steps, which is the main contribution of this study. In conclusion, future directions and possible improvements for Industry 4.0 are briefly given.

1.2   Main Concepts and Components of Industry 4.0

In recent years, Industry 4.0 has attracted great attention from both manufacturing companies and service systems. On the other hand, there is no certain definition of Industry 4.0 and naturally, there is no definite utilization of the emerging technologies to initiate the transformation of Industry 4.0. Mainly, Industry 4.0 is comprised of the integration of production facilities, supply chains and service systems to enable the establishment of value added networks. Thus, emerging technologies such as big data analytics, autonomous (adaptive) robots, cyber physical infrastructure, simulation, horizontal and vertical integration, Industrial Internet, cloud systems, additive manufacturing and augmented reality are necessary for a successful adaptation. The most important point is the widespread usage of Industrial Internet and alternative connections that ensure the networking of dispersed devices. As a consequence of the developments in Industrial Internet, in other words Industrial Internet of Things, distributed systems such as wireless sensor networks, cloud systems, embedded systems, autonomous robots and additive manufacturing have been connected to each other. Additionally, adaptive robots and cyber physical systems provide an integrated, computer-based environment that should be supported by simulation and three-dimensional (3D) visualization and printing. Above all, entire system must involve data analytics and miscellaneous coordination tools to conduct a real time decision making and autonomy for manufacturing and service processes.

While constructing the framework, network of sensors, real-time processing tools, role-based and autonomous devices are interpenetrated with each other for real-time collection of manufacturing and service system data. In order to understand the proposed framework which is addressed in this study, this section gives detailed information about supportive technologies and design principles underlined for Industry 4.0 implementation with real life cases and examples. After that, proposed framework is presented with regard to design principles and supportive technologies for acquiring context-aware operational system including smart products and smart processes.

1.2.1 State of Art

For successful system adaptation to Industry 4.0, three features should be taken into account: (1) horizontal integration via value chains, (2) vertical integration and networking of manufacturing or service systems, and (3) end to-end engineering of the overall value chain (Wang et al. 2016). Vertical integration requires the intelligent cross-linking and digitalization of business units in different hierarchal levels within the organization. Therefore, vertical integration enables preferably transformation to smart factory in a highly flexible way and provides the production of small lot sizes and more customized products with acceptable levels of profitability. For instance, smart machines create a self-automated ecosystem that can be dynamically subordinated to affect the production of different product types; and a huge amount of data is processed to operate the manufacturing processes easily. On the other hand, horizontal integration obtains entire value creation between organizations for enriching product life cycle using information systems, efficient financial management and material flow (Acatech 2015). The horizontal and vertical integration enable real time data sharing, productivity in resource allocation, coherent working business units and accurate planning which is crucial for connected devices in the term, Industry 4.0. Finally, end-to-end engineering assists product development processes by digital integration of supportive technologies considering customer requirements, product design, maintenance, and recycling (Wang et al. 2016).

1.2.2 Supportive Technologies

For successful implementation of Industry 4.0 transformation, three core and nine fundamental technologies are required to be the part of the entire system. In this section, detailed information of these supportive technologies is given for better understanding of the proposed framework.

Adaptive robotics: As a consequence of the combination of microprocessors and AI methodologies, the products, machines and services become smarter in terms of having not only the abilities of computing, communication, and control, but also having autonomy and sociality. In this regard, adaptive and flexible robots combined with the usage of artificial intelligence provide easier manufacturing of different products by recognizing the lower segments of each parts. This segmentation proposes to provide decreasing production costs, reducing production time and waiting time in operations. Additionally, adaptive robots are useful in manufacturing systems especially in design, manufacturing and assembly phases (Wittenberg 2015). For instance, assigned tasks are divided into simpler sub problems and then are constituted a set of modules in order to solve each sub problem. At the end of each sub task completion, integration of the modules to reach an optimal solution is essential. One of the sub technologies underlying adaptive robots can be given from co-evolutionary robots that are energetically autonomous and have scenario based thinking and reaction focused working principle (Wang et al. 2016).

A real life example can be given: a robot called Yumi which is created for ABB manufacturing operations. Yumi has flexible handling, parts-feeding mechanism, camera based part location detection system and state-of-the-art motion control for the adaptation of ABB production processes as reported in ABB Contact (2014). Another example can be given as Kuka KR Quantec robot that has task-distributing screws and other production material by delivering the ordered KANBAN boxes coming from the central warehouse rack. The "workerbot", created from pi4, has a humanoid anatomy with two arms, a rotating upper body and supported by camera and image processing systems. This combined mechanism enables memory based activity identification using independent recognition of the previous positions and characteristics of production parts (VDMA 2016).

The general characteristics of these applications are given in the following:

- Networked via Ethernet or Wi-Fi for high speed data transmission
- Easy integration in existing machinery communication systems
- Optical and image processing of part positioning
- Integrated robot controller
- Memory based or case based learning mechanism.

Embedded systems (*Cyber physical infrastructure*): Embedded systems, named as Cyber-Physical Systems (CPS), can be explained as supportive technology for the organization and coordination of networking systems between its physical infrastructure and computational capabilities. In this respect, physical and digital tools should be integrated and connected with other devices in order to achieve decentralized actions. In other words, embedded systems generally integrate physical reality with respect to innovative functionalities including computing and communication infrastructure (Bagheri et al. 2015).

In general, an embedded system obtains two main functional requirements: (1) the advanced level of networking to provide both real-time data processing from the physical infrastructure and information feedback from the digital structure; and (2) intelligent data processing, decision-making and computational capability that support the physical infrastructure (Lee et al. 2015). For this purpose, embedded systems consist of RTLS technologies, sensors, actuators, controllers and net-worked system that data or information is being transformed and transferred from every device. In addition to that, information acquisition can be derived from data processing and data acquisition in terms of applying computational intelligence supported by learning strategies such as case based reasoning.

A specific example for embedded systems can be observed in Beckhoff main-tenance tool: Process parameters (stress, productive time etc.) of mechanical components can be recorded digitally while making some adjustments such as technical experiments in online or offline platforms. In addition to that case, cyber-physical research and learning platform "CP Factory" from Festo provides educational institutions and companies with access to the technology and appli-cations of Industry 4.0. The main part of the (physical) mechanism is supported by an intelligent module for the communication of process data—the "CPS Gate". The "CPS Gate" operates within the factory's workstations as the "backbone" module for controlling the processes. Schunk linear motor drives with each prioritized order in the assembly lines repeatedly for decentralized quality assurance and docu-mentation of quality criteria (VDMA 2016).

The embedded systems have some properties mentioned as follows:

- Increased operational safety through the detection of safety-critical status prior to their importance level,
- Sensorless or with sensor switching condition monitoring,
- Control and monitoring using feedback loops,
- Systematical and targeted integration of storage and analysis of data directly and interactively on the local control, in private networks or in the public cloud system,
- Flexible and reconfigurable parts and machines.

Additive manufacturing: Additive manufacturing is a set of emerging tech-nologies that produces three dimensional objects directly from digital models through an additive process, particularly by storing and joining the products with proper polymers, ceramics, or metals. In details, additive manufacturing is initiated by forming computer-aided design (CAD) and modeling that arranges a set of digital features of the product and submit descriptions of the items to industrial machines. The machines perform the transmitted descriptions as blueprints to form the item by adding material layers. The layers, which are measured in microns, are added by numerous of times until a three-dimensional object arises. Raw materials can be in the form of a liquid, powder, or sheet and are especially comprised of plastics, other polymers, metals, or ceramics (Gaub 2015). In this respect, additive manufacturing is comprised of two levels as software of obtaining 3D objects and material acquisition side.

Although barriers to the existing technology are appeared especially in production processes, there are incomparable properties using 3D printers and additive manufacturing. For instance, additive manufacturing processes outperform than conventional manufacturing mechanisms for some products including shaping initially impossible geometries such as pyramidal lattice truss structures. Obviously, printing mechanism reduces material waste by utilizing only the required materials (Ford 2014). Besides that, networked system comprised of ordering, selection of injection molding is also necessary to monitor the process variables and parameters on a particular interface. Customer requirements are also involved in the manufacturing design and necessary components for these plastic parts' manufacturing are gathered in advance. The injection molding machine encapsulates the metal blades and the information system for design features interconnects the individual design process steps with proper additive manufacturing system operations. In addition to that, a laser-marking phase is also adopted in the production line (Gaub 2015).

Real life example is aroused from ARBURG GmbH that deals with individualized high volume plastic products. An ALLROUNDER injection moulding machine and a freeformer for additive manufacturing are linked by means of a seven-axis robot to 3D plastic lettering using additive processes (VDMA 2016).

Cloud technologies: Cloud based operating is another essential topic for the contribution of networked system integration in Industry 4.0 transformation. The term "cloud" includes both cloud computing and cloud based manufacturing and design. Cloud manufacturing implies the coordinated and linked production that stands "available on-demand" manufacturing. Demand based manufacturing uses the collection of distributed manufacturing resources to create and operate reconfigurable cyber-physical manufacturing processes. Here, main purpose is enhancing efficiency by reducing product lifecycle costs, and enabling the optimal resource utilization by coping with variable-demand customer focused works (Thames and Schaefer 2017a, b). Comprehensively, cloud based design and manufacturing operations indicate integrated and collective product development models based on open innovation via social networking and crowd-sourcing platforms (Thames and Schaefer 2017a, b).

As a consequence of the advancements in cloud technologies such as decreasing amount of reaction times, manufacturing data will increasingly be practiced in the cloud systems that provide more data-driven decision making for both service and production systems (Rüßmann et al. 2015). On the other hand, according to "From Industry 4.0 to Digitizing Manufacturing" report submitted by Manufacturing Technology Center, privacy and security issues aroused from system lacks are needed to be considered and secondly, extra storage needs, payment options and physical location should be carefully decided in advance (Wu et al. 2014). At the same time, productivity increases in advance: an example is from GE Digital that proposed "Brilliant Manufacturing Suite" which uses smart analytics to evaluate operational data and factory's overall equipment effectiveness is increased by 20% or more. Besides that, M&M Software's industrial cloud service platform is based on real time data analytics and consists of a universal core system of individual web portals. The mentioned system can be remotely operated on both a PC using a browser and on mobile devices.

The requirements of cloud based processing are listed as follows:

- Data driven applications are worked on cloud-based infrastructure, and every supply chain element and user is connected through the cloud system.
- Real time data analytics for notifications and abnormalities using independent cloud database function.
- Take full advantage of big data to optimize system performance according to external and sudden changes.
- Users need a connected device to see the necessary information on cloud, and they have authorized access to available applications and data worldwide.
- Proactive application function as an automatic shift log or tool change log, perform adaptive feed control, detect collisions, monitor processes, and much more besides.

Virtualization technologies (Virtual Reality (VR) and Augmented Reality (AR)): Virtualization technologies are based on AR and VR tools that are entitled the integration of computer-supported reflection of a real-world environment with additional and valuable information (Paelke 2014). In other words, virtual information can be encompassed to real world presentation with the aim of enriching human's perception of reality with augmented objects and elements (Syberfeldt et al. 2016). For this purpose, existing VR and AR applications associate graphical interfaces with user's view of current environment. The essential role of graphical user interfaces is that users can directly affect visual representations of elements by using commands on appeared on the screen and interacts with these menus referenced by ad hoc feedbacks.

According to these purposes, visualization technologies have four functional requirements: (i) scene capturing, (ii) scene identification, (iii) scene processing, (iv) scene visualization. Thus, hardware such as handheld devices, stationary visualization systems, spatial visualization systems, head mounted displays, smart glasses and smart lenses are utilized for implementation. On the other hand, key challenges for the adaptation of visualization cases present the environment with realistic objects for better user experience, adding necessary information via meta graphics and enriching users' perception by color saturation and contrast. With this respect, approaches for visualization technologies' displays are based on three focuses: (i) video-based adaptation supported by the camera that assists augmented information, (ii) optical adaptation that user gives information by wearing a special display and (iii) projection of stated objects (Paelke 2014).

Today, visualization technologies are mainly applied in diversified fields such as video gaming, tourism and recently, this topic has started to be considered within the context of constructing quality management systems, assembly line planning and organizing logistics and supply chain actions for smart factories (Paelke 2014; Syberfeldt et al. 2016; DHL report). Specific examples can be given from BMW Connected Drive that enables navigation information and assists driver assistance systems, Q-Warrior helmet for military purposes, Liver explorer for medical practitioners and Recon Jet for leisure activities (DHL Report 2015). Particularly, AR and VR systems are adapted to computer aided quality assessment for the estimation of scale, tracking the product position and visualizing current state of the product by a graphical user interface. In shop floor implementation of visualization technologies, video based glasses (Oculus Rift), optical glasses (C wear) and Android based devices, video based tablet and spatial projector are utilized. Final example could be given for logistics, especially considering warehouse operations, transportation optimization, last mile delivery, customer services and maintenance. In this virtual world, operators can interact with machines or other devices by using them on a cyber-representation and change parameters in order to interpret the operational and maintenance instructions (Segovia et al. 2015). The most remarkable future implementation of visualization systems is the requirement of tailor made solutions for human and robot collaboration and more user-friendly devices for better experience (Rüßmann et al. 2015).

The visualization technologies have some properties mentioned as follows:

- Optimal user support through augmented reality and gamification.
- Significantly more convenient and user-friendly interface design.
- The mobile projection providing holistic and latency-free support.

Simulation: Before the application of a new paradigm, system should be tested and reflections should be carefully considered. Thus, diversified types of simulation including discrete event and 3D motion simulation can be performed in various cases to improve the product or process planning (Kühn 2006). For example, simulation can be adapted in product development, test and optimization, production process development and optimization and facility design and improvement. Another example could be given from Biegelbauer et al.'s (2004) study that handles assembly line balancing and machining planning that requires to calculate operating cycle times of robots and enables design and manufacturing concurrency.

In the perspective of Industry 4.0, simulation can be evaluated as a supportive tool to follow the reflections gathered from various parameter changes and enables the visualization in decision-making. Therefore, simulation tools can be used with other fundamental technologies of Industry 4.0. For instance, simulation based CAD integration ensures the working of multiple and dissimilar CAD systems by changing critical parameters. Additionally, simulation can reflect what-if scenarios to improve the robustness of processes. Especially for smart factories, virtual simulation enables the evaluation of autonomous planning rules in accordance with system robustness (Tideman et al. 2008).

Data Analytics and Artificial Intelligence: In consequence of the manufacturing companies start to adopt advanced information and knowledge technologies to facilitate their information flow, a huge amount of real-time data related to manufacturing is accumulated from multiple sources. The collected data which is occurred during R&D, production, operations and maintenance processes is increasing at exponential speed (Zhang et al. 2016). In particular, data integration and processing in Industry 4.0 is applied for improving an easy and highly scalable adaptation for dataflow-based performance analysis of networked machines and processes (Blanchet et al. 2014). Data appears in large volume, needed to be processed quickly and requires the combination of various data sources in diversified formats. For instance, data mining techniques have to be used where data is gathered from various sensors. This information assists the evaluation of current state and configuration of different machinery, environmental and other counterpart conditions that can affect the production as seen in smart factories. The analysis of all such data may bring significant competitive advantage to the companies that they are able to be meaningfully evaluate the entire processes (Obitko and Jirkovský 2015).

Some of the data mining approaches combined with support vector machines, decision tree algorithm, neural networks, heuristic algorithms are successfully applied for clustering classification and deep learning cases. Additionally, data mining approaches are generally combined with operation research methods including mixed integer programming and stochastic programming. For instance, data visualization problems caused by high dimensional data are especially faced in big data management and to overcome this problem, adaption of quadratic assignment problem formulations is required in advance.

Unlike data processing in relational databases, three functions should be considered in order to build big data infrastructure that can operate successfully with Industry 4.0 components: (i) Big data acquisition and integration (ii) Big data processing and storage (iii) Big data mining and knowledge discovery in database. Big data acquisition and integration phase includes data gathering from RFID readers, smart sensors and RFID tags etc. Big data processing and storage configures real time and non-real time data as a form of structured and unstructured data by cleaning, transforming and integration. Finally, big data mining is adopted by clustering, classification, association and prediction using decision trees, genetic algorithm, support vector machines and rough set theory for big data mining and knowledge discovery. Particularly, big data mining does not only necessitate a certain understanding of the right application but also requires dealing with unstructured data. Thus, huge amount of data preparation including specifying substantial variables and extracting appropriate data are conducted for making precise prediction and classification (Zhang et al. 2016).

Communication and Networking (Industrial Internet): Communication and networking can be described as a link between physical and distributed systems that are individually defined. Using communication tools and devices, machines can interact to achieve given targets, focus on embedding intelligent sensors in real-world environments and processes. Industrial Internet of Things (IIoT) relies on both smart objects and smart networks and also enables physical objects integration to the network in manufacturing and service processes. In other words, major aim of IIoT is to provide computers and machines to see and sense the real world applications that can provide connectivity from anytime, anywhere for anyone for anything (IERC 2011).

The main requirements for communication and networking are listed as (i) distributed computing and parallel computing for data processing, (ii) Internet Protocol (IP), (iii) communication technology, (iv) embedded devices including RFID tags or Wireless Sensor Networks (WSN) and (v) application, (Borgia 2014). In addition to these requirements, Uckelmann et al. (2011) added Internet of People and Intranet/Extranet of Things to reflect the integrity of interior parts of business in them and enhance service orientation with effective contiguity of other devices. On the other hand, the main issue for the integration period is the construction of standards for the communication of various devices. Companies also face another problem, security flaws, as realized from privacy issues (Zuehlke 2010).

Thanks to the recent advances of decreasing costs for sensor networks, NFC, RFID and wireless technologies, communication and networking used for IIoT suddenly became an engaging topic for industry and end-users. The potential of IIoT is significant: it is predicted that the number of IIoT will reach a potential by 50 billion in 2020, which demonstrates the scalability of IoT (Qin et al. 2016). The determination of the physical status of objects through sensors and integration of Web 2.0 technologies can cause the huge collection and processing of operational data, allows real time response as a reaction of the status of things (IERC 2011). Today, interoperability with big data processing platforms can provide with agent based services, real-time analytics, and business intelligence systems which is essential for networking.

Considering manufacturing advancements supported by communication and networking technologies, manufacturing industries are ready to improve the production processes with big data analytics to take the advantage of higher compute performance with open standards and achieve the availability of industry know-how in advance (Pittaway et al. 2004). As a result of the penetration of manufacturing intelligence, manufacturers can be able to enhance quality, increase manufacturing output. This knowledge provides better insights for detecting root cause of production problems and defect mapping, monitor machine performance and reduce machine failure and downtime. Therefore, IIoT or communicative systems are not only considered as a technology of Industry 4.0 but also evaluated as a "cover" that contains many features from Industry 4.0 tools (IERC 2011). An example could be given form predictive maintenance: Liggan and Lyons (2011) indicated that a sustainable predictive maintenance includes the mechanical evaluation of the production processes such as motor rating, number of pumps and valves, belt length, thermal imaging and base vibration analysis. Thus, the integrated system should process the data by considering the historical data captured from sensors and other environmental conditions such as material quality, comments of the material gathered from other users. The collection can be supported by using Web 2.0 technologies and extracting knowledge from the collected data can be transformed to organizational "know how. This process requires the assistance of big data analytics and obviously, real time tracking system should be implemented considering two ways: (i) data collection (ii) ordering to the machines or services using knowledge emerged from big data analysis. For this reason, communication and networking can be evaluated as an inclusive technology that support the functioning of other Industry 4.0 tools such as big data analytics, simulation and embedded systems.

Thus far, we discussed supportive technologies for Industry 4.0 in details. These technologies require a fundamental structure for the successful implementation. Therefore, RTLS and RFID technologies, cyber security, sensors and actuators and mobile technologies are the infrastructure for supportive technologies.

RTLS and RFID technologies: Smart Factory has some critical operations such as smart logistics, transportation and storage by satisfying efficient coordination of embedded systems and information logistics. These operations include identification, location detection and condition monitoring of objects and resources within the organization and across company using Auto-ID technologies. The aggregation and processing of the real time data gathered from production processes and various environmental resources assist the integration of organization functions and enables self-decision making of the machines and other smart devices. Thus, radio-frequency identification (RFID) and real time location systems (RTLS) may generate value in manufacturing and logistics operations as Uckelmann (2008) described the basic concepts of real time monitoring systems in the following way:

• Identification—especially RFID with single and bulk reading,
• Locating—RTLS like GPS and others,
• Sensing—e.g. temperature and humidity sensors.

In this respect, the possibility of item based tracking—for logistics processes (e.g. control of incoming goods) and also essential for production processes (e.g. control of correct parts assembled)- ensures the automation of the existing processes and remanufacturing of parts. Thus, practitioners widely adapt RTLS and RFID based technologies for successful implementation of smart factories and processes. For instance, Hologram Company RAKO GmbH implemented a smart identification label that enables electronic identification of the individualized products easily and reliably either on the product itself or on the packaging. The tags used in HP-digital machinery plant have unique serial numbers such as data matrix code, QR code or standard barcode. Another example can be given from advanced assembly line for floor cleaning machines of Alfred Kärcher GmbH that QR code embedded with a RFID chip is utilized to track the product from the beginning of the production. In this case, data is read out at every workstation in order to follow detailed assembly instructions appeared on a monitor at a specific workstation (VDMA 2016).

The outcomes of RTLS and RFID based systems are appeared as follows:

• Process-optimized production of a product in a large number of versions
• Enhanced functionality and flexibility of the assembly line
• A high degree of data transparency

- Real time data flow to enable rapid support for workers.

Cyber security: As mentioned in previous sections, Industry 4.0 transformation requires intensive data gathering and processing activities. Thus, security of the data storage and transfer processes is fundamental concepts for companies. The security should be provided in both cloud technologies, machines, robots and automated systems considering the following issues:

- Data exportation technologies' security
- Privacy regulations and standardization of communication protocols
- Personal authorization level for information sharing
- Detection and reaction to unexpected changes and unauthorized access by standardized algorithms.

To avoid the results of these issues, operational recovery, end user education, network security and information security should be ensured by cyber incident response, critical operation recovery and authorization level detection programs. Other preventive actions can be access controls of user account, firewalls, intrusion detection systems and penetration tests that use the vulnerability scanners.

A real life example can be given as CodeMeter from Wibu-Systems AG that IP protection mechanisms prevent illegal copying and reverse engineering of software, theft of production data, and product counterfeiting; machine code integrity foils tampering, Freud detection and cyber-attack identification. In addition to that, a hidden counter sitting inside the software license controls volume productions, making sure only the identified batches are produced. The entire cyber security system provides remote communication adapted by using certificate chains and combined with digital signature and assists end point security for sensors, devices, and machines.

Benefits of cyber security systems are given as follows:

- Encryption algorithms for hardware-based protection,
- Trustworthy communication protocols between sensors, devices, and machines enabled by using digital signature and certificates,
- Flexible licensing models and authorization level detection,
- Faster back office automation with the seamless integration of licenses in all leading CRM, ERP, and e-commerce systems (VDMA 2016).

Sensors and actuators: Sensors and actuators are the basic technology for embedded systems as entire system obtains a control unit, usually one or more microcontroller(s), which monitors(s) the sensors and actuators that are necessary to interact with the real world. In industrial adaptation of Industry 4.0, embedded systems similarly consist of a control unit, several sensors and actuators, which are connected to the control unit via field buses. The control unit conducts signal processing function in such systems. As smart sensors and actuators have been developed for industrial conditions, sensors handle the processing of the signal and the actuators independently check production current status, and correct it, if necessary. These sensors transmit their data to a central control unit, e.g. via field buses. In this respect, sensors and actuators can be defined as the core elements for entire embedded systems (Jazdi 2014).

An example of the adaptation of sensors and actuators to Industry 4.0 implementation can be given from in intelligent pneumatics actualized from AVENTICS GmbH. In this case, Advanced Valve (AV) series are adapted with pneumatic valves, sensors, or actuators connected to the valve electronics. This connection links the embedded system to higher-level control by the adaptation of IIoT. The AES supports all conventional fieldbuses and Ethernet protocols for a seamless flow of data to implement preventive maintenance. Another example can be given from Bosch GmbH that the system enables monitoring product quality in supply chain. In that case, transport packaging is furnished with integrated Bosch sensors that are connected to the Bosch IoT cloud system. They continuously record data that are relevant for product quality, such as temperature, shocks or humidity.

The benefits of sensors and actuators are:

• Real-time tracking along the entire production or service systems
• Continuous documentation and data collection for supporting big data analytics, deep learning and knowledge extraction
• Enriched system availability via condition monitoring (VDMA 2016).

Mobile technologies: Mobile devices made a significant progress after these devices were first introduced and are now so much more than just basic communication tools. These devices ensure the internet enabled receiving and processing of large amounts of information and are provided with high quality cameras and microphones, which again allow them to record and transmit information.

Considering the implementation of communication and networking in Industry 4.0 adaptation, connectivity to inanimate objects allows companies to communicate with each other. When mobile devices become internet enabled and enriched by Wi-Fi technology, they come to the same platform as other process equipment does. This situation implies that mobile devices can receive and transmit process related data in advance, and allow users to address issues as they cope with in real time decision making. Using mobile technologies, issues can now be recognized and dealt with faster as information moves with a higher velocity in the right position. The mobile devices are now used in practical way and able to interact with process equipment, material, finished goods and parts through IIoT.

Before implementing Industry 4.0, design principles should be taken into account. The design principles provide the comprehensive adaptation of entire system and enable the coordination between Industry 4.0 components, which are discussed in the following part. There are seven design principles appeared in the application and implementation of Industry 4.0: Agility, Interoperability, Virtualization, Decentralization, Real-time data management, Service Orientation and Integrated business processes. Interoperability implies the communication of cyber physical systems components with each other using Industrial Internet and regular standardization processes to create a smart factory. Additionally, virtualization enables monitoring of entire system, new system adaptation and system changes using simulation tools or augmented reality. Decentralization is a key term for self-decision making of the machines and relies on the learning from the previous events and actions. Real time data management is the tracing and tracking the system by online monitoring to prevent system lacks when a failure appears. Service orientation is the satisfaction of customer requirements adaptation to entire system with using a perspective of integrating both internal and external sub systems. Integrated business process is the link between physical systems and software platforms by enabling communication and coordination mechanism assisted by corporate data management services and connected networks. Last principle, agility means the flexibility of the system to changing requirements by replacing or improving separated modules based on standardized software and hardware interfaces (Hermann et al. 2015). Considering these principles, academicians can search further implementations and frameworks for Industry 4.0 and practitioners would be able to implement Industry 4.0 components to the autonomous system properly.

1.3   Proposed Framework for Industry 4.0

The main motivation of Industry 4.0 is the connection and integration of manufacturing and service systems to provide effectiveness, adaptability, cooperation, coordination and efficiency, as realized from design principles (Li et al. 2015a, b). Therefore, correlation between design principles and existing technologies is explained in Table 1.1 for better understanding of proposed framework.

According to Table 1.1, interoperability of communicative components could be satisfied using cyber physical system security and Industrial Internet of Things adaptation such as communication and networking. In similar manner, monitoring the changes in existing system can be provided by simulation modeling and virtualization techniques such as augmented reality and virtual reality. An example could be given from CAutoD which optimizes the existing design process of trial-and-error by altering the design problem to a simulation problem, as an automating digital prototyping. Additionally, adaptive robots, embedded systems based on cyber physical infrastructure, cloud systems and big data analytics should be successfully combined in order to enable self-decision making and autonomy. For instance, by utilizing data processing, analysis and sharing, knowledge discovery can be extracted and preventive actions can be ensured for each cyber physical component. RFID and RTLS technologies, sensors, and actuators are the major components for real time data management in terms of traceability and real time reaction to sudden changes appeared in sub systems. To illustrate, real time maintenance systems can set a precedent by presenting the integration of real time data processing. By this way, possible preventive precautions are taken via RTLS platforms and sensors against instantaneous incidents. Cloud systems, data analytics and artificial intelligence techniques also ensure the specific customer specifications by assessing the external information from digital manufacturing environment and fulfill service-oriented architecture of Industry 4.0 framework.

Table 1.1 Categorization of Industry 4.0 technologies and design principles

Technologies	Principles						
	(1) Real time data management (collection/processing/analysis/inference)	(2) Interoperability	(3) Virtualization	(4) Decentralization	(5) Agility	(6) Service orientation	(7) Integrated business processes
Adaptive robotics					●		
Data analytics & Artificial intelligence	●			●	●	●	
Simulation			●		●		
Embedded systems				●			
Communication & Networking		●		●	●		●
Cyber security		●					●
Cloud technologies					●	●	●
Additive manufacturing					●		
Virtualization technologies (VR & AR)			●		●		
Sensors & Actuators	●			●	●		●
RFID & RTLS technologies	●			●	●		●
Mobile technologies					●		

Considering the coordination between design principles and supportive technologies, agility and integrated business processes can be evaluated as the most important design principles. In this regard, integrated business process implies the relation between cyber security and cloud systems that are based on a communication and networking infrastructure such as Industrial Internet. Besides that, connected and networked adaptive robots, additive manufacturing, cloud systems, data analytics and artificial intelligence play an important role for the adaptation to changing requirements to satisfy system stability and agility. For instance, data acquisition about breakdowns gathered from data analytics studies and knowledge transfer via cloud systems enable "learning" factory. Here, 3D printing is an inevitable tool for delivering compatible parts in shortest time before production operations are disturbed by breakdowns (European Commission Report 2015; Saldivar et al. 2015).

Considering the reflections gathered from the relationship between design principles and supporting technologies, a general framework for Industry 4.0 adaptation is presented as seen in Fig. 1.1. To enable a successful implementation of Industry 4.0, companies should focus on involving and redefining the smart product and smart processes to their core functions such as product development, manufacturing, logistics, marketing, sales and after sale services. In this respect, a smart product

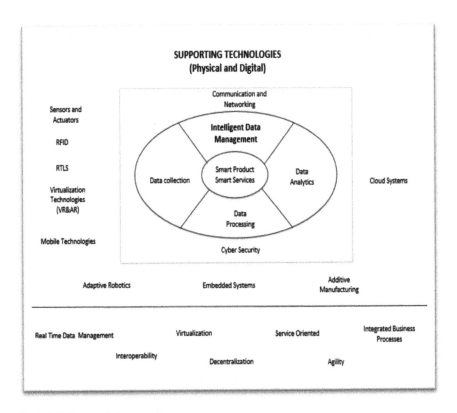

Fig. 1.1 Industry 4.0 framework

contains three basic components: (i) physical part(s) including a mechanical part, (ii) a smart part that has sensors, microprocessors, embedded operating system and user interface (iii) connectivity part that has ports, antenna and protocols (Porter and Heppelmann 2015). All smart products and processes should have an entire supporting technology platform that relies on the connection and coordination of data exchange, data collection, data processing and analytics between the product and services to external sources. Using big data analytics, products and services can be monitored and changes can be observed in numerous environmental conditions. Additionally, cloud technologies ensure coordinated and linked production to distributed systems. As a consequence, interoperability with big data processing platforms are strengthened by agent-based services, real-time analytics, and business intelligence systems, which is essential for networking. Thus, big data platforms and cloud systems can provide real time data management in order to give fast reactions for data processing, management of data flow and extracting know how to improve entire product performance and utilization. In this way, adjustments can be made according to difference between current condition and desired requirements by adapting algorithms and iterative processes such as self-learning and self-assessment. This intelligent data management should be promoted by the construction of communication and networking infrastructure based on Industrial Internet and cyber security for successful remote controlling and monitoring.

The core technologies that underlined for supporting technologies are sensors and actuators, RFID and RTLS technologies, virtualization technologies and mobile technologies. To satisfy virtualization part of Industry 4.0, augmented reality and Virtual reality are inevitable tools. Virtual reality (VR) provides a computer-aided simulation tool for reflecting the recreation of real life environment that user feels and sees the simulated reality as they are experiencing in real life. On the other hand, augmented reality is progressed in applications to combine digital elements with real world actions. The overall integration of VR and AR provides the enrichment of real life cases and actions. Furthermore, RFID, sensors and RTLS technologies enable real time data flow and data gathering, which is essential for intelligent data management in decentralized systems. Additionally, mobile technologies enable receiving and processing of large amounts of information to record and transmit information and supports agile-remote control of entire business.

Supporting technologies such as adaptive robotics, embedded systems and additive manufacturing can be established based on the core technologies. For instance, embedded systems are constructed on the integration of physical systems including sensors and actuators to enhance the autonomous nature of Industry 4.0. Besides that, additive manufacturing enables digital models through an additive process by shaping 3D features for agile manufacturing. Thus, core technologies' adaptation should be appropriately provided before implementing supporting technologies.

Because of the improvements in supporting technologies, new business models, remote services and continuous production operations are aroused. For instance, many companies are initiated to offer their products as a service that enhances win-win strategy for both companies and customers. Additionally, companies mainly focus on the entire systems, not the single components of the systems separately. Here, main question is, will the industrial companies focus on closely linked products or create a platform that satisfies overall related products? Besides these questions, continuous production operations imply the interconnected products and processes that utilize networking and cloud technologies. In fact, product is actually a proposed technology or a platform that requires the sustainability of product life cycle. Here, the critical issues are being a part of shared responsibility for cyber security and participating the standardization processes to assure regular data organization and sharing.

1.4   Conclusion

Interconnected and smart products are dramatically adapted for value creation in manufacturing and other areas after rapid changes appeared in the combination of manufacturing and computer technology. As a result of Industry 4.0 indicates more productive and continuous systems, companies have been searching the right adaptation of this term. This situation necessitates the clear explanation of the implementation strategy. Thus, in this study, we first focused on the explanation of the core and supportive technologies and description of the design principles for better understanding of the proposed framework. After that, link between design principles and technologies are described in details. Finally, conceptual framework for Industry 4.0 is proposed by concerning fundamental links between smart products and smart processes.

From the experiences of industrial companies, future directions indicate that production, control and monitoring of the smart and connected products will change from human labor centered production to fully automated way. In this respect, transformation of Industry 4.0 requires strategic work force planning, constructing right organization structure developing partnerships and participating and sharing the technological standardization, which are essential factors to drive technological advancements. As realized from McKinsey's report (2016), major implementation areas in manufacturing will be real time supply chain optimization, human robot collaboration, smart energy consumption, digital performance management and predictive maintenance. Additionally, supportive technologies will be more effective by the adaptation of nanotechnology and robotics to Industry 4.0 implementation. Moreover, self-organized, self-motivated and self-learning systems will be experienced by using more sophisticated artificial intelligence algorithms and auto-creation of the business processes will be encountered in near future.

References

ABB Contact (2014) Connecting the world—Industry 4.0, Customer news magazine by ABB

Acatech (2015) Final report of the Industrie 4.0 Working Group. Technical Report. http://www.gtai.de/GTAI/Content/EN/Invest/_SharedDocs/Downloads/GTAI/Brochures/Industries/industrie4.0-smart-manufacturing-for-the-future-en.pdf

Bagheri B, Yang S, Kao H, Lee J (2015) Cyber-physical systems architecture for self-aware machines in Industry 4.0 environment. IFAC-Pap Online 48–3:1622–1627

Biegelbauer G, Vincze M, Noehmayer H, Eberst C (2004) Sensor based robotics for fully automated inspection of bores at low volume high variant parts. IEEE International Conference on Robotics and Automation, vol 5, pp 4852–4857, 26 April–1 May 2004

Bitkom, VDMA, ZVEI (2016) Implementation strategy Industrie 4.0: report on the results of the Industrie 4.0 Platform

Blanchet M, Rinn T, von Thaden G, de Thieulloy G (2014) Industry 4.0: the new industrial revolution. How Europe will succeed. In: Think Act, Roland Berger Strategy Consultants GmbH

Borgia E (2014) The internet of things vision: key features, applications and open issues. Comput Commun 54:1–31

Deloitte AG (2015) Industry 4.0: challengers and solutions for the digital transformation and use of exponential technologies

DHL Report (2015) Internet of things in logistics: a collaborative report by DHL and Cisco on implications and use cases for the logistics industry. http://www.dpdhl.com/content/dam/dpdhl/presse/pdf/2015/DHLTrendReport_Internet_of_things.pdf

European Commission (2015) Factories of the Future. http://ec.europa.eu/research/industrial_technologies/factories-of-thefuture_en.html. Viewed 12 May 2017

Ford SLN (2014) Additive manufacturing technology: potential implications for U.S. manufacturing competitiveness. J Int Commer Econ. http://www.usitc.gov/journals

Gaub H (2015) Customization of mass-produced parts by combining injection molding and additive manufacturing with Industry 4.0 technologies, reinforced plastics, vol 00, Number 00. doi:10.1016/j.repl.2015.09.004

Hermann M, Pentek T, Otto B (2015) Design principles for Industrie 4.0 scenarios: a literature review, Technische Universität Dortmund, Fakultät Maschinenbau, Audi Stiftungslehrstuhl Supply Net Order Management, White Paper

IERC (2011) Internet of things: strategic research roadmap [WWW Document]. URL: http://www.internet-of-things-research.eu/about_iot.htm. Accessed 26 April 17

Jazdi N (2014) Cyber physical systems in the context of Industry 4.0. 2014 IEEE international conference on automation, quality and testing, robotics. doi:10.1109/AQTR.2014.6857843

Kühn W (2006) Digital factory—simulation enhancing the product and production engineering process. In: Perrone LF, Wieland FP, Liu J, Lawson BG, Nicol DM, Fujimoto RM (eds) Proceedings of the 2006 winter simulation conference, pp 1899–1906

Lasi H, Fettke P, Feld T, Hoffmann M (2014) Industry 4.0. Bus Inf Syst Eng 6(4):239–242

Lee J, Bagheri B, Kao H (2015) A cyber-physical systems architecture for Industry 4.0-based manufacturing systems. Manufact Lett 3:18–23

Li X, Li D, Wan J, Vasilakos AV, Lai CF, Wang S (2015a) A review of industrial wireless networks in the context of Industry 4.0. Wirel Netw. doi:10.1007/s11276-015-1133-7

Li S, Xu LD, Zhao S (2015b) The internet of things: a survey. Inf Syst Front 17:243–259

Liggan P, Lyons D (2011) Applying predictive maintenance techniques to utility systems. Pharm Eng 31(6):1–7

McKinsey Digital (2016) Industry 4.0 after the initial hype: where manufacturers are finding value and how they can best capture it

Obitko, M, Jirkovský V (2015) Big data semantics in Industry 4.0. In: Mařík V et al (eds) HoloMAS 2015, LNAI 9266, pp 217–229

Paelke V (2014) Augmented reality in the smart factory supporting workers in an Industry 4.0. Environment. IEEE Emerging Technology and Factory Automation (ETFA)

Pittaway L, Robertson M, Munir K, Denyer D, Neely A (2004) Networking and innova tion: a systematic review of the evidence. Int J Manag Rev 5:137–168

Porter M, Heppelmann J (2015) How smart connected products are transforming companies. Harvard Business School Publishing

Qin J, Liu Y, Grosvenor R (2016) A categorical framework of manufacturing for Industry 4.0 and beyond. Procedia CIRP 52:173–178

Rüßmann M, Lorenz M, Gerbert P, Waldner M, Justus J, Engel P, Harnisch M (2015) Industry 4.0: the future of productivity and growth in manufacturing industries. https://www. bcgperspectives.com/content/articles/engineered_products_project_business_industry_40_ future_productivity_growth_manufacturing_industries/

Saldivar AAF, Li Y, Chen W, Zhan Z, Zhang J, Chen LY (2015) Industry 4.0 with cyber-physical integration: a design and manufacture perspective. Proceedings of the 21st international conference on automation & computing, Glasgow, UK, September 2015

Segovia D, Mendoza M, Mendoza E, González E (2015) Augmented reality as a tool for production and quality monitoring. 2015 international conference on virtual and augmented reality in education. Procedia Comput Sci 75:291–300

Syberfeldt A, Holm M, Danielsson O, Wang L, Brewster RL (2016) Support systems on the industrial shop-floors of the future—operators' perspective on augmented reality. 6th CIRP Conference on Assembly Technologies and Systems (CATS). Procedia CIRP 44, pp 108–113

Thames JL, Schaefer D (2017) Cybersecurity for Industry 4.0 and advanced manufacturing environments with ensemble intelligence. In: Thames L, Schaefer D (eds) Cybersecurity for Industry 4.0.1. Springer (Springer Series in Advanced Manufacturing), Berlin, pp 243–265

Thames JL, Schaefer D (2017) Industry 4.0: an overview of key benefits, technologies, and challenges. In: Thames, L, Schaefer D (eds) Cybersecurity for Industry 4.0.1. Springer (Springer Series in Advanced Manufacturing), Berlin, pp 1–33

Tideman M, van der Voort MC, van Houten FJAM (2008) A new product design method based on virtual reality, gaming and scenarios. Int J Interact Des Manuf (IJIDeM) 2(4):195–205. doi:10. 1007/s12008-008-0049-1

Uckelmann D, Harrison M, Michahelles F (2011) An architectural approach towards the future internet of things. In: Uckelmann D et al (eds) Architecting the internet of things. doi:10.1007/ 978-3-642-19157-2_1

Uckelmann D (2008) A definition approach to smart logistics. In: Balandin S, Moltchanov D, Koucheryavy Y (eds) Next generation teletraffic and wired/wireless advanced networking. NEW2AN 2008. Lecture Notes in Computer Science, vol 5174. Springer, Berlin

VDMA (2016) Industrie 4.0 in practice—Solutions for industrial applications. Industry 4.0 Forum

Wang L, Wang G (2016) Big data in cyber-physical systems, digital manufacturing and Industry 4.0. Int J Eng Manufact 4: 1–8

Wang S, Wan J, Zhang D, Li D, Zhang C (2016) Towards smart factory for Industry 4.0: a self-organized multi-agent system with big data based feedback and coordination. Comput Netw 000:1–11

Wittenberg C (2015) Cause the trend Industry 4.0 in the automated industry to new requirements on user interface. In: Kurosu M (ed) Human-computer interaction, Part III, HCII 2015, LNCS, vol 9171, pp 238–245

Wu D, Rosen DW, Schaefer D (2014) Cloud-based design and manufacturing: status and promise. In: Schaefer, D (ed) Cloud-Based Design and Manufacturing (CBDM). Springer, Berlin, 1–24

Zhang Y, Ren S, Liu Y, Si S (2016) Big data analytics architecture for cleaner manufacturing and maintenance processes of complex products. J Cleaner Prod Volume 142 Part 2, 1–16

Zuehlke D (2010) Smart factory towards a factory-of-things. Ann Rev Control 34(1): 129–138. [Online]. Available: http://www.sciencedirect.com/science/article/pii/S1367578810000143

Chapter 2
Smart and Connected Product Business Models

Sezi Cevik Onar and Alp Ustundag

Abstract A business model describes the value offered by the company. Business models have a significant impact on the success of the business. Smart and connected products, which connect the physical objects by using sensors and communication technology, change the nature of traditional businesses and business models. The value propositions, revenue streams, and technologies offered with these smart and connected products are different from the traditional business models. In this chapter, we define the key features of smart and connected product business models and reveal the successful real life cases with this framework.

2.1 Introduction

The fourth industrial revolution is realized by the combination of numerous physical and digital technologies such as sensors, embedded systems, cloud computing and Internet of Things (IoT). Regardless of the triggering technologies, the main purpose of industrial transformation is to increase the resource efficiency and productivity to increase the competitive power of the companies. The transformation era, which we are living in now, differs from the others in that it not only provides the change in main business processes but also reveals the concepts of smart and connected products by presenting service-driven business models.

According to Osterwalder et al. (2005), the business model is a description of the value a company offers to one or several segments of customers and of the architecture of the firm and its network of partners for creating, marketing, delivering this value and relationship capital, and generating profitable and sustainable revenue streams. A successful business model should have four fundamental building

S. Cevik Onar (✉) · A. Ustundag
Department of Industrial Engineering, Faculty of Management,
Istanbul Technical University, 34367 Macka, Istanbul, Turkey
e-mail: cevikse@itu.edu.tr

A. Ustundag
e-mail: ustundaga@itu.edu.tr

© Springer International Publishing Switzerland 2018
A. Ustundag and E. Cevikcan, *Industry 4.0: Managing The Digital Transformation*,
Springer Series in Advanced Manufacturing, https://doi.org/10.1007/978-3-319-57870-5_2

blocks (1) the customer value proposition that fulfills an important job; (2) the profit formula that lays out how your company makes money delivering the value proposition; (3) the key resources that value proposition requires; and (4) the key processes needed to deliver it (Innosight 2017). In the harsh competition of today's business world, only the companies developing the right business models can be successful. The companies having innovative business models can transform businesses, create new markets and unlock significant growth.

With the rise of IoT, the growing number of smart and connected products entered into the market, change industry domains and the structure of competition. With service-driven business models, they are reshaping industry boundaries and creating entirely new industries (Porter and Heppelman 2014). Smart, connected product capabilities can be grouped into four categories (PTC 2017):

- Monitor: Sensors and external data sources enable monitoring of the product's condition, operation, and external environment to generate alerts and actionable intelligence.
- Control: Software built into the product enable control and personalization.
- Optimize: Monitoring and controlling capabilities enable optimization algorithms to enhance product performance and perform remote service and repair.
- Automate: Combination of monitoring, controlling and optimization capabilities enhanced with software algorithms and business logic allows the product to perform autonomously.

Smart and connected products having enhanced capabilities allow the radical change in business models. A shift from a product-based to service-centric business models has emerged (Porter and Heppelman 2014). This transformation forces the companies to differentiate their value chain alignment, set new strategic decisions to cope with competition, redefine the organizational structure and change their application success factors.

Value propositions, revenue streams, and technologies are the primary determinants of smart and connected product business models. In this chapter, we propose a business model framework for smart and connected products. In this framework, the main determinants are classified with relevant cases.

The rest of the chapter is organized as follows: Sect. 2.2 briefly explains the nature of business models. In Sect. 2.3, the key business model components of smart and connected products are given. The proposed framework is provided in Sect. 2.4. Conclusions and further suggestions are given in the last section.

2.2 Business Models

Business modeling is a useful tool that makes current processes in the system less costly, more efficient and satisfying profit expectation with excellent progress. Business modeling is determining a company's priority value about customer expectation, developing methods by forming foundations about determined priority and progress of working on providing continuity to these methods. As many entrepreneur acts could not have satisfied the expectation of current era, they were not able to stay alive. Entrepreneurship is meaningful when it fulfills a necessity. Modeling this need on a particular value provides continuity to a company. Businesses that do not progress with specific methods and systems have limited life, as they cannot take work under control in today's world. According to scientific researches, there are various business models. However, as each company has developed its model on its priority, no particular model is applicable for every company. This project involves information about existing models and progress of applying existing models. A deep research about models shows that they provide holistic approaches to businesses and reveal the research needs in details.

In literature, different studies focus on various values. Some authors have supported the idea of focusing on customer demands, and some supported the idea of focusing on determination of the company. A business model is a comprehensive tool to understand the way of doing the business of the firms and to analyze their performance and competitive strategies through the design of their products or services offered to the market. Business models also help for a full scanning of a firm. With a business model, a person can understand a company's costs to produce, the value they offered to the market and which strategy they follow to communicate with the consumers. The business models are the value propositions that explains how the company meets the customer needs (Kim and Mauborgne 2005). A business model is a system of interdependent activities that enlarge the boundaries of an organization (Amit and Zott 2001). The companies should realize customer expectations and they should build up a new business model structure based on the value created.

Weisbord (1976) defines purpose, structure, relationship, reward, leadership, helpful mechanisms as the key components of a business model. According to Amit and Zott (2001), the blue ocean strategies and business models can be classified as across alternative industries, strategic groups and chain of buyers, complementary offerings, emotional and functional appeal. The value proposition, value creation, delivery, and value capture are considered as the key components of a business model. In the value proposition, the company decides which product/service they offer to their consumers and which customer segment they have. The last thing to do in the first part is choosing the way of communicating with the customers. The company analyzes what they have for these customer segments and how they implement their business regarding their resources, channels, partners and technology. After all these processes, in the last part, the company should introduce its cost structure and determine its revenue streams to defray this spending.

Osterwalder and Pigneur (2010) synthesize the business models in the literature and develop a comprehensive template for business models. In this model, there are five key components of a business structure, infrastructure, offering, customers, finances, and resources. The beginning of building the model should be with Customer Segments part. Before analyzing their strategy and activities, companies should define which customer segments, they target. They should make a segmentation of the customers according to their importance and then choose the best target group for the following steps. The next step should be valued proposition. Each firm should focus on value, which affects the customers to improve a new business strategy. This value should be determined according to the understanding of the client expectations. After detecting the value proposition, the company should decide which channels they use to provide these benefits for the customers. The next step will be customer relationships. In this part, the company should define their way of communicating with their customer. The company should decide, whether it be a face-to-face meeting or via calling and so on. After evaluating the customer parts, the company should define the revenue streams, which shows the things, or activities, which will make money. The next stage should be detecting the key partners. During this business, the company can need to do same partnerships with other companies or some foundations. Companies should define these partners. After these parts, the critical part called key activities should be fill in. Next step is setting action plans to reach this customer segment with supplying particular value via detected channels to gain the aimed revenue stream. After a well-designed action plan, the company should control its key resources if it is enough for the short and long-term plans or not. The last stage is determining cost structure, which is the total spending of this model.

Successful smart and connected product business models differentiate from the traditional business models regarding the value propositions, revenue streams, and the technologies.

2.3 Key Business Model Components of Smart and Connected Products

Business models are the management tools that facilitate creating, enlarging and retaining business value. In the recent years, there is a significant interest in smart and connected business models, especially IoT business models, since they enhance competitive advantage (Wirtz et al. 2016). There are many challenges of these business models. For instance, connecting different devices and developing standards or maintaining information security are some of the challenges of smart and connected business models (Hognelid and Kalling 2015). Business models are the tools that help to overcome these difficulties. In this section, the key components of smart and connected business models are examined.

Guo et al. (2017) evaluate the impact of the business models of an IoT on the business value. They classified business models as novelty-centered, efficiency-centered, lock-in centered and complementary. Novelty centered IoT business models focus on creating new markets, new services or innovations. Efficiency focused IoT business models try to increase the efficiency of transactions. The objective of these models is to fasten, simplify, eliminate errors and improve the transparency of a transaction. Lock-in centered IoT business models try to enlarge transaction volume and increase customer loyalty by various ways such as customizing, improving safety and reliability. Complementary based IoT business models provide additional goods/services that are more valuable together. Bujari et al. (2017) reveal the capabilities and limitations IoT technologies. IoT business models along with some factors such as security and privacy are considered as the factors that affect IoT usage. Weinberger et al. (2016) claim that IoT enhances industries and supply chains by shortening optimization cycles, testing processes and increasing the quality, flexibility, and efficiency. The IoT business models are classified into six groups namely, remote usage and condition monitoring, object self-service, digital add-on, digital lock-in, product as a point of sales and physical freemium.

Dijkman et al. (2015) adopted business canvas model to the IoT business models. In this study, key partners in IoT models are hardware producers, software developers, and other suppliers, data integration, launching customers, distributors, logistics and service partners. The main activities are customer development, product development, implementation/service, marketing/sales, platform development, software development, partner management, logistics. Key resources are physical resources, intellectual property, employee capabilities, financial resources, software, and relations. Value propositions are newness, performance, customization, getting the job done, design, brand/status, price, cost reduction, risk mitigation, accessibility, convenience/usability, comfort, and the possibility for updates. Customer relationship components are personal assistance, dedicated assistance, self-service, automated service, communities, and co-creation. Channels are sales-force, web sales, own stores, partner stores, and wholesaler. Customer segments are mass market, niche market, segmented, diversified and multi-sided platforms. Cost structures are product development cost, IT cost, personnel cost, hardware/production cost, logistics cost, marketing, and sales cost. Asset sale, usage, rental, subscription, licensing, installation and advertising fees are the main revenue streams.

2.4 Proposed Framework

The smart and connected business models can be classified based on value propositions, revenue stream, and the offered architecture and technologies. Value proposition categorizes the values created with business models. Revenue stream shows how the business model creates income and the used technologies can be explained under three layers, namely, the physical layer, connectivity layer, and digital layer. Combinations of these futures create benefits for the customers.

2.4.1 Value Proposition

The smart and connected product technologies significantly transform the core business models. They not only provide cost reductions but also create new revenue streams. These business models offer four primary values. In this chapter, the value propositions of smart and connected product businesses are classified based on Guo et al.'s (2017) classification as follows:

- Novelty

Novelty refers to the new market, new services and innovation. IoT applications enable firms creating new markets, new services, and innovations. Not only the IoT solutions that enable creating new business models but also the IoT Platforms themselves can be considered as novel business models.

Watson IoT Platform enables manufacturers to develop personalized adaptive robots. Similarly, Libelium provides an IoT Platform that allows connecting various devices located at different locations and gather data from these devices. Cities, agriculture systems or water resources can be managed by using Libelium platform. Microsoft IoT Platform provides a range of solutions to connect, analyze and optimize the usage of industrial equipment and devices in the factories. This platform can significantly enhance the workplace safety.

AT&T Connected car turns the vehicle into a Wi-Fi hotspot. Keeps the car connected to a network even during the trip. Users can do video streaming, web browsing, and Internet radio via Car's network.

- Efficiency

The smart and connected product business models that bring efficiency make the transaction faster, simple, transparent and eliminate errors. The primary goal of smart and connected product applications can be increasing the efficiency of transactions. The efficiency can be improved by making the transaction faster, simple or increasing its dependability and transparency. Increasing efficiency usually the first step of implementing embedded systems. Therefore, efficiency is the most common benefit sought in IoT businesses.

For instance, IoT embedded vibration sensors developed by Intel can track the vibration in bridges and provide reliable data for maintenance planning which may increase the reliability of constructions. Nest Thermostat developed by Nest Labs tracks user behaviors and decreases heating and cooling costs by adjusting the desired room temperature before the owner arrives at home. It also connects with the energy companies and enables these companies to optimize their production levels based on the consumption information.

- Lock-in

Smart and connected product applications allow firms to enlarge transaction volume for existing customers and increase customer loyalty. They can provide affiliate programs or virtual communities. These applications can enhance customization, repeat usage, customer retention, reliability, and transaction safety.

Amazon Dash Button is a Wi-Fi connected device that reorders product with the press of a button (https://www.amazon.com/Dash-Buttons). These dash buttons are paired with pre-selected products. Pushing the button is enough for a reorder. This smart and connected product enables repeat usage of the pre-selected items.

Sentrian (Platform) is a remote patient monitoring platform analyzes bio-sensor data and sends patient-specific alerts to clinicians and learn from their feedbacks.

- Complementary

Complementary based IoT business models provide additional goods/services that are more valuable together. Cross selling and effective bundling are available with these business models.

Amazon Dash Replenishment Service enables automatic reordering for products such as laundry detergent for washing machines or ink cartridges for printers. These embedded replenishment services give automatic orders before the last load. It uses the Amazon's customer service systems such as payment systems and authentication.

Similarly, Moov is a fitness wearable with an artificial intelligent based personal coach.

2.4.2 IoT Value Creation Layers and Technologies

The IoT is a widely used term for a set of technologies, systems, and design principles associated with the emerging wave of Internet-connected things that are based on the physical environment (Holler et al. 2014). An IoT architecture consists of three main layers which can be called as physical, connectivity and digital layers (Andre 2015). In this section, Streetline Smart Parking Soluiton will be used as an example to explain how the value creation layers are formed. Streetline is a smart parking company using IoT technologies which help to solve parking issues by delivering smart data and advanced analytics to its customers (Streetline 2017).

- Physical Layer

At the physical layer, sensors and micro-controllers work together to provide one of the most important aspects of the Internet of Things (IoT): Detecting changes in an object or the environment, allowing for capture of relevant data for real-time or post-processing. Sensors are used for detection of physical changes including temperature, light, pressure, sound, and motion. They are also used for detection of the logical relationship of one object to another(s) and the environment including the presence/absence of an electronically traceable entity, location or activity (Research and Markets Report 2016). Actuators are other critical devices at the physical layer which are used to effect a change in the environment such as the temperature controller of an air conditioner.

As an example in Streetline Smart Parking Solution at the physical layer, ultra-low power sensors detect the status of parking spaces and communicate through a wireless mesh network. The mesh network is built using a series of unobtrusive and easily installed repeaters, placed on locations like streetlights and telephone poles, to form a canopy above the targeted area. Meter Monitors can be installed in legacy single-space meters to detect payment and wirelessly integrate the meter into the network (Streetline and IBM White Paper).

- Connectivity Layer

The connectivity layer is responsible for connecting to other smart things, network devices, and servers. Its features are also used for transmitting and processing sensor data (Sethi and Sarangi 2017). IoT devices connect and communicate using various technical communication models and technologies such as IP networks, 3G/4G, Bluetooth, Z-Wave, WiFi ZigBee, RFID or NFC (Internet Society IoT Report 2015).

- The device-to-device communication model represents two or more devices that directly connect and communicate between one another, rather than through an intermediary application server.
- In a device-to-cloud communication model, the IoT device connects directly to an Internet cloud service like an application service provider to exchange data and control message traffic.
- In the device-to-gateway model, or more typically, the device-to-application-layer gateway (ALG) model, the IoT device connects through an ALG service as a conduit to reach a cloud service.
- The back-end data-sharing model refers to a communication architecture that enables users to export and analyze smart object data from a cloud service in combination with data from other sources.

In Streetline Smart Parking Solution, the sensors are tied into municipality Wi-Fi networks using a device-to-gateway model at the connectivity layer.

- Digital Layer

Digital layer stores, analyzes and processes huge amounts of data that comes from the connectivity layer. It can manage and provide a diverse set of services to the lower layers. It employs many technologies such as databases, cloud computing, and big data processing modules. It is also responsible for delivering application specific services to the user. It defines various applications in which the IoT can be deployed, for example, smart homes, smart cities, and smart health (Sethi and Sarangi 2017).

In Streetline Smart Parking Solution, at the digital layer, machine-learning techniques are deployed to merge multiple data sources into an integrated data set for real-time parking guidance. The data analytics capabilities are used to improve accuracy and provide a comprehensive city-wide view of parking utilization. A suite of mobile and web applications enables each key stakeholder in the parking ecosystem.

Table 2.1 shows the characteristics some smart and connected business model cases. Both businesses to business (B2B) and business to customers (B2C) business models are considered in this table.

Table 2.1 Characteristics some smart and connected business model cases

Company/Product name	Business	Value proposition	Value proposition	Revenue stream	Physical layer	Connectivity layer	Digital layer
IBM/Streetline	B2C	Efficiency	Streetline fastens finding parking spots by giving directions to the open parking slots	Parking payment can be done via mobile app, it has a monthly subscription model, parking enterprises can publish their parking slots	Sensors with light sensitivity installed in the parking slots	The sensors are tied into municipality Wi-Fi networks. Data obtained from sensors are transferred to cloud platform	Data are analyzed at cloud analytic platform, and customers use consumer-facing streetline parker app
Nestlab	B2C	Efficiency	Nest thermostat developed by Nest Labs tracks user behaviors and decreases heating and cooling costs by adjusting the desired room temperature before the owner arrives at home. It also connects with the energy companies and enables these companies to optimize their production levels based on the consumption information	Nest sells its smart home consumer products, namely its thermostat, which is more expensive than a traditional thermostat. It sells the energy consumption data that it is able to aggregate across its devices	It has a split-spectrum smoke sensor that detects carbon monoxide, a microphone that enables complete self-monitoring and other sensors that detect temperature, humidity, occupancy and ambient light	The sensors are tied into Wi-Fi networks	Nest app enables remote usage of products
Amazon's dash button	B2C	Lock-in	Amazon's dash button is a Wi-Fi connected device that reorders	Dash buttons are sold at low prices	Sensors and Wi-Fi connection are	The sensors are tied into Wi-Fi networks	Dash button is managed with a mobile app

(continued)

Table 2.1 (continued)

Company/Product name	Business	Value proposition	Value proposition	Revenue stream	Physical layer	Connectivity layer	Digital layer
			product with the press of a button. These dash buttons are paired with pre-selected products		embedded in the dash buttons		
Amazon's dash replenishment	B2B/B2C	Complementary	Amazon dash replenishment service enables automatic reordering for products such as laundry detergent for washing machines	Companies can use amazon's authentication and payment systems, customer service, and fulfillment network for automatic reordering	It works with any sensor or tracking mechanism, but it can tracks customer usage directly	The device is connected to the internet through companies cloud or directly	Dash replenishment service uses Login with Amazon (LWA), Amazon Simple Notification Service, and RESTful API endpoints to allow the device or cloud to integrate
AT&T/Connected car	B2C	Novelty	Turns vehicle into a Wi-Fi hotspot. Keeps the car connected to a network even during the trip. Users can do video streaming, web browsing, and internet radio via car's network	No initial payment required. Monthly payment with a 2-year contract. Payment changes with the content of the internet package	A plug-in, which is attached to the car, provides Wi-Fi connection to the people in the car	Plug-in is connected to AT&T's 4G LTE network.	People in the car can connect to the internet from their smart devices via Wi-Fi network in their car. Also, they can locate their car in the parking lot, send directions to their car and remote start through the mobile app
Shipwise	B2B	Efficiency	Shipwise provides a web platform for small and medium enterprises who want	Shipwise makes contract based arrangements small	All the equipping vehicles, warehouses, goods, cargo carriers and other devices in	The sensors are tied into Wi-Fi networks	Shipwise's online dashboard provides shippers end-to-end

(continued)

Table 2.1 (continued)

Company/Product name	Business	Value proposition	Value proposition	Revenue stream	Physical layer	Connectivity layer	Digital layer
			to book and manage global freight shipments. Shipwise's online dashboard enables tracking global shipments and provides analyzing tools to reduce errors	and medium enterprises	the supply chain have embedded smart sensors		management of shipments
Placemeter	B2B	Efficiency	Placemeter turns video into meaningful data. It provides solutions for smart cities, out-of-home advertising, and retailers	Monthly subscription model with a 1-year contract, which includes a cellular data plan, full feature access, software updates and 24/7 technical assistance	The platform collects videos through Arlo cameras or existing security cameras	Security cameras are connected to closed or hidden Wi-Fi networks and to the cloud server	From Placemeter's dashboard, users can access all the information about the movement in that specific location
Belkin/WeMo	B2C	Efficiency	A Smart plug that is controllable from smartphones. Power on/off situation can be arranged according to a predetermined schedule. Also, tracks energy consumption of the related electronic device. It decreases the energy consumption of that plug	One time payment model. Customers have to pay a certain price for the purchase phase. No additional payment during the usage	WeMo is plugged into an electrical outlet, and it controls power outage according to the signals coming from the Wi-Fi network	WeMo is tied into Wi-Fi networks. Which then connects to a cloud server	Via the cloud server, users can control their electronic products which are connected to WeMo from the mobile app

(continued)

Table 2.1 (continued)

Company/Product name	Business	Value proposition	Value proposition	Revenue stream	Physical layer	Connectivity layer	Digital layer
Philips/Hue	B2C	Efficiency	A light bulb that can be turned off and on remotely from a mobile device. It also can change the color of the light	One-time payment model. Customers have to pay a certain price for the purchase phase. No additional payment during the usage	The small internal device provides communication with the network	Through WiFi network reception of bytes are transmitted to Hue. It has three communication channels: router, Ethernet, and ZigBee	With Philips Hue's mobile app, users send commands to the light bulb
August/Smart lock	B2C	Efficiency	A smart lock that unlocks when the owner of the house came home with his smartphone and locks when the owner leaves. Access right can be granted to others via application in the smartphone	One time payment model. Customers have to pay a certain price for the purchase phase. No additional payment during the usage	Command signals through Wi-Fi or Bluetooth are received by the receptors	August smart lock is tied into Wi-Fi networks through august connect. Which then connects to a cloud server	Via mobile application, which is connected to the lock through Wi-Fi network, users can control august smart lock
DHL/IoT tracking and monitoring	B2B	Efficiency	Implementation of IoT applications to the logistics regarding providing tracking and monitoring situation of packages or the vehicles, at the same time giving information about the	Decreases the cost of monitoring	Location and utilization of machines or vehicles are determined through signals coming from the sensors	The sensors are tied into Wi-Fi networks. Data obtained from sensors are transferred to cloud platform	Data are analyzed at cloud analytic platform, and customers reach information via mobile DHL's mobile app

(continued)

Table 2.1 (continued)

Company/Product name	Business	Value proposition	Value proposition	Revenue stream	Physical layer	Connectivity layer	Digital layer
			maintenance status. It decreases the costs related to logistics				
Farmobile	B2C	Efficiency	Provides farmers to collect machine and agricultural data. From the mobile app digital information can be tracked. Efficiency can be increased by sharing the agricultural data with the agronomists and insurance agents	The yearly subscription model, which includes a cellular data plan, full feature access, unlimited user accounts, unlimited data storage, software updates and 24/7 technical assistance	With small devices (sensors) called PUCs, machinery information is collected	The sensors are tied into Wi-Fi networks. Data obtained from sensors are transferred to cloud platform second by second	Data are analyzed at cloud analytic platform, and customers use consumer-facing farmobile app
Condeco sense	B2B	Efficiency	Workplace occupancy sensor that gathers real-time data from workplaces. Gathered data shows typically the utilization rate of the desk or workplace areas like meeting rooms	One time payment model. Customers have to pay a certain price for each unit of products. No additional payment during the usage	Wireless sensors collect real-time data about workplace occupancy. Each sensor monitors movement and heat in that particular workplace	The sensors are tied into Wi-Fi networks. Data obtained from sensors are transferred to cloud platform	The collected raw data is then analyzed and transformed into insights to help companies manage the office are more effectively from the smart devices
Daqri/Smart helmet	B2B	Effectiveness	Increases the capabilities in the workplace via data visualization, thermal vision, guided work	One time payment model which include: Daqri helmet, Compute pack with Daqri VOS installed, Developer Tools,	High-speed wide angle camera tracking and collecting the visual information around. RGB camera, stereo infrared cameras, and	The cameras are connected to Intel Core processor	The processor provides high-performance multimedia and Augmented Reality applications. Data visualization, thermal

(continued)

Table 2.1 (continued)

Company/Product name	Business	Value proposition	Value proposition	Revenue stream	Physical layer	Connectivity layer	Digital layer
			instruction, and remote expert function	Updates, and device management for enterprises	infrared light projector. Also, has a thermal camera		vision, and work instructions are provided to the user in the workplace
Spire	B2C	Efficiency	Tracks users breath and activity, and warns in the stress situations to calm down the user. Even the user do not feel tense, Spire can detect it and warn the user to get calm via sending notifications to users smart phone	One time payment model. Customers have to pay a certain price for the purchase phase. No additional payment during the usage	Spire's custom force sensors detect the reflection of breathing to torso and diaphragm. Then analyzes the breath periods to understand the user's mood	Spire uses a very popular and low power wireless protocol called Bluetooth to send and receive data to and from your phone	Analyzed data at Spire's processor is sent to Spire's mobile app, which user can track its breath data
Semios	B2B	Efficiency	Semios provides farmers to manage and protect their orchard. Frost, leaf wetness, soil moisture and pest pressure can be controlled and viewed real-time	The yearly subscription model, which includes a cellular data plan, full feature access, unlimited user accounts, unlimited data storage, software updates and 24/7 technical assistance	Semios sensors in the orchard collect all the data about the field like pests, weather conditions and irrigation	The sensors are directly connected with semios platform through cell providers	Through the Semios app, users can get notifications, reach actionable intelligence and see the records
SkyBitz/Tank monitoring	B2B	Efficiency	Provides remote monitoring and analytics for storing liquids and compressed gases like	The monthly subscription model, which includes monitor, sensor and the cable	Through the sensors inside of the tanks, tank monitoring collects information of	The sensors are tied into Wi-Fi networks. Data obtained from sensors are	Data are analyzed at Skybitz Tank Monitoring Portal which users can reach 24/7 and see all the

(continued)

Table 2.1 (continued)

Company/Product name	Business	Value proposition	Value proposition	Revenue stream	Physical layer	Connectivity layer	Digital layer
			petroleum, chemicals or water. Optimizes operation efficiency with wireless monitoring		tank level, pressure level, and location	transferred to cloud platform	information about their tank
Enevo	B2B	Efficiency	Enevo helps municipalities or enterprises to simply their waste and providing an effective solution for waste management	The service is provided with 1–10 years of contract with a subscription fee	Enevo, waste management analytics, uses advanced ultrasonic sonar technology to detect fill levels of waste bins	The sensors are tied into Wi-Fi networks. Data obtained from sensors are transferred to cloud platform	At Enevo platform users can manage their waste and reach to all information about it
Wellintel	B2B	Efficiency	It is a groundwater information system for homeowners, farmers, and scientists. It helps to track groundwater level and sends alerts in the necessary situations	One-time paying model, which includes a cellular data plan, full feature access, unlimited user accounts, unlimited data storage, software updates and 24/7 technical assistance	Sensors that are placed in the well collects all information required about the well	The sensors are tied into Wi-Fi networks. Data obtained from sensors are transferred to cloud platform	Data are analyzed at WellIntel platform which users can reach 24/7 and see all the information about their tank
Garagio	B2C	Efficiency	It provides users to control their garage door from anywhere, anytime	One-time paying model, which includes a cellular data plan, full feature access, unlimited user accounts, unlimited data storage, software updates and 24/7 technical assistance	The black box that is inserted into garage communicates with the garage door and transfers the data to Wi-Fi network	The black box is tied into home Wi-Fi networks. Data obtained from the black box are transferred to cloud platform	The web and app-based Garagio dashboard enable users to view garagedoors activity, share access and control the garagedoor

2.5 Conclusion and Further Suggestions

This chapter presents research that leads to a framework for smart and connected product business model. Using an enhanced literature review, we define the main determinants of a business model and specific types these main determinants. Subsequently, we classified the smart and connected product cases based on the main determinants.

In the literature, few studies that focus on smart and connected product business models. This chapter fulfills this need by defining the key features, value propositions, revenue streams, and technologies. This chapter can guide future smart and connected business models.

Although our study yields new and insightful results for smart and connected product business models, a more comprehensive approach can be applied to see the different applications in various markets. A qualitative study can be implemented to show the correlations among the key determinants.

References

Amit R, Zott C (2001) Value creation in e-business. Strateg Manag J 22:493–520
Andre P (2015). http://labs.sogeti.com/iot-connect-physical-and-digital-worlds-for-new-business-models/
Bujari A, Furini M, Mandreoli F, Martoglia R, Montangero M, Ronzani D (2017) Standards, security and business models: key challenges for the IoT scenario mobile networks and applications, pp 1–8 (Article in Press)
Dijkman RM, Sprenkels B, Peeters T, Janssen A (2015) Business models for internet of things. Int J Inf Manage 35:672–678
Guo L, Wei S-Y, Sharma R, Rong K (2017) Investigating e-business models' value retention for start-ups: the moderating role of venture capital investment intensity. Int J Prod Econ 186:33–45
Hognelid P, Kalling T (2015) Internet of things and business models. Proceedings of the 9th international conference on standardization and innovation in information technology, IEEE SIIT 2015, art. no. 7535598
Holler J, Tsiatsis V, Mulligan C, Avesand S, Karnouskos S, Boyle D (2014) From machine-to-machine to the internet of things: introduction to a new age of intelligence. Elsevier Academic Press, Cambridge
Innosight (2017). https://www.innosight.com/leadership-agenda/innovating-business-models/
IOT Report—Internet Society (2015). https://www.internetsociety.org/sites/default/files/ISOC-IoT-Overview-20151014_0.pd
Kim CW, Mauborgne R (2005) Blue ocean strategy: how to create uncontested market space and make the competition irrelevant. Harvard Business School Press, Boston
Osterwalder A, Pigneur Y (2010) Business model generation: a handbook for visionaries, game changers, and challengers, New Jersey, USA: John Wiley & Sons Inc, ISBN: 978-0-470-87641-1
Osterwalder A, Pigneur Y, Tucci CL (2005) Clarifying business models: origins, present, and future of the concept. Commun Assoc Inf Syst 16(1):1
Porter ME, Heppelmann JE (2014) How smart, connected products are transforming competition. Harvard Bus Rev 65–88

PTC Smart Connected Products (2017). Retrieved from: http://support.ptc.com/WCMS/files/160474/en/PTC_eBook_Impact_of_the_IoT_on_Manufacturers.pdf

Research and Markets Report (2016). Retrieved from: http://www.prnewswire.com/news-releases/sensors-and-embedded-systems-in-the-internet-of-things-iot-2016-2021-market-analysis-and-forecasts–for-the-16-trillion-business—research-and-markets-300319924.html

Sethi P, Sarangi SR (2017) Internet of things: architectures, protocols, and applications. J Elect Comput Eng art. no. 9324035 p 25 10.1155/2017/9324035

Streetline (2017). https://www.streetline.com/company/

Streetline—Leading the way for 'Smart Parking' HBR. https://rctom.hbs.org/submission/streetline-leading-the-way-for-smart-parking/

Streetline and IBM White Paper. https://www-01.ibm.com/common/ssi/cgi-bin/ssialias?htmlfid=GVS03037USEN

Weinberger M, Bilgeri D, Fleisch E (2016) IoT business models in an industrial context. At-Automatisierungstechnik 64(9): 699–706

Weisbord Marvin R (1976) Organizational diagnosis: six places to look for trouble with or without a theory. Group & Organization Studies 1(4):430–447

Wirtz BW, Pistoia A, Ullrich S, Göttel V (2016) Business models: origin, development and future research perspectives. Long Range Plann 49(1) pp 36–54, ISSN 0024-6301, http://dx.doi.org/10.1016/j.lrp.2015.04.001

Chapter 3
Lean Production Systems for Industry 4.0

**Sule Satoglu, Alp Ustundag, Emre Cevikcan
and Mehmet Bulent Durmusoglu**

Abstract Succeeding a cultural and people oriented transformation, Lean Producers adopts the philosophy of doing more with less by eliminating non-value-added activities from production processes to maintain effectiveness, flexibility, and profitability. With the context of Industry 4.0, new solutions are available for combining automation technology with Lean Production. Moreover, when effective resource (finance, labour, material, machine/equipment) usage is concerned, it is obvious that Industry 4.0 should be applied on lean processes. In this context, this chapter attempts to emphasize the interaction between Lean Production and Industry 4.0 and proposes a methodology which provides guidance for Industry 4.0 under lean production environment. Moreover, Industry 4.0 technologies and automation oriented Lean Production applications are also included.

3.1 Introduction

"Out of clutter, find simplicity. From discord, find harmony. In the middle of difficulty lies opportunity."

Albert Einstein

Evolved from the conceptualization of Toyota Production System by Taichii Ohno's initiatives at Toyota Motor Company, Lean Manufacturing can be descri-

S. Satoglu (✉) · A. Ustundag · E. Cevikcan · M.B. Durmusoglu
Faculty of Management, Department of Industrial Engineering,
Istanbul Technical University, Macka, 34367 Istanbul, Turkey
e-mail: onbaslis@itu.edu.tr

A. Ustundag
e-mail: ustundaga@itu.edu.tr

E. Cevikcan
e-mail: cevikcan@itu.edu.tr

M.B. Durmusoglu
e-mail: durmusoglum@itu.edu.tr

© Springer International Publishing Switzerland 2018
A. Ustundag and E. Cevikcan, *Industry 4.0: Managing The Digital Transformation*,
Springer Series in Advanced Manufacturing, https://doi.org/10.1007/978-3-319-57870-5_3

bed as a multi-faceted production approach comprising a variety of industrial practices, directed towards identifying value adding processes from the purview of customer and to enable flow of these processes at the pull of the customer through the organization (Sanders et al. 2016). Currently, it is being adopted by many companies in the aftermath of the 1973 oil shock. Notable rewards have been reported by European firms through this effort, not only in manufacturing sectors, but also in service fields such as retail, healthcare, travel and financial services (Piercy and Rich 2009). The main purpose of the system is to eliminate through improvement activities various kinds of waste lying concealed within a system. Even during periods of slow growth, Toyota could make a profit by decreasing costs through a production system that completely eliminated excessive inventory and work force. It would probably not be overstating the case to say that this is another revolutionary production management system which follows the Taylor system (scientific management) and Ford system (mass-assembly line). Companies applying Lean Manufacturing tools ultimately want to meet customer demands with fewer resources and less waste.

From technological point of view, Lean Production can be regarded as a complement to automation. Zuehlke (2010) suggested that the complexity of the production systems should be reduced by lean practices and expressed that relying too much on technology cannot always improve the performance but may make the system more complicated. Lean Production suggests five different levels of automation which should be considered when deciding appropriate automation strategy as shown in Table 3.1 (Rother and Harris 2001). In the first level, everything is done manually. The operator loads the machine and starts the machine, and the machine cycles. Next the operator unloads the part and manually transfers it to the next production step. The second level of automation is when the operator manually loads the machine, the machine automatically cycles, and the operator manually removes the part and takes it to the next station. In level three automation, the operator manually loads the part into the machine, and the part automatically cycles. The part is automatically unloaded from the machine and the operator then moves the part. Level four automation automatically loads the part; it's automatically cycled, automatically unloaded, and then manually transferred to the next process. Finally, level five automation is entirely automatic. The machine is automatically loaded, cycled, and unloaded, and the part is transferred by automation. Meanwhile,

Table 3.1 Levels of Lean Automation

Level	Loading machine	Machine cycle	Unloading machine	Transferring part
1	Operator	Operator	Operator	Operator
2	Operator	Automatic	Operator	Operator
3	Operator	Automatic	Automatic	Operator
THE GREAT DIVIDE (Cost and flexibility changes drastically.)				
4	Automatic	Automatic	Automatic	Operator
5	Automatic	Automatic	Automatic	Automatic

there is a great divide between level three and level four automation. When making the jump to level four automation, cost (maintenance, engineering, machine etc.) often increases while flexibility can decrease (Harris and Harris 2008).

In fact, applications for the integration of Industry 4.0 into Lean Production already exist and has been expressed as the term Lean Automation. Lean Automation aims for higher changeability and shorter information flows to meet future market demands. Contrary to popular belief, Lean Production does not exclude automation. According to its autonomation principle of Lean Production, repeating and value-adding tasks should be automated (Bilberg and Hadar 2012). With the term Low Cost Intelligent Automation (LCIA), it is suggested that standardized, automated, flexible and cost-efficient solutions should be favoured over customised solutions (Takeda 2006). However, LCIA only focuses on mechanical and electrical systems and does not consider information and communication technology. Both Lean Production and Industry 4.0 favour decentralised structures over large, complex systems and both aim for small, easy to integrate modules with the low level of complexity (Takeda 2006; Kolberg et al. 2016).

Taking into account the above-mentioned manners, this chapter emphasizes the relationship between Industry 4.0 and lean manufacturing, and proposes a methodology providing a waste hunting environment for Industry 4.0 applications. This research also gives professionals the insight of looking for every opportunity to improve effectiveness while committing to Industry 4.0. The rest of the chapter is organized as follows. In Sect. 3.2, relevant literature is reviewed. Proposed methodology is presented in Sect. 3.3. Cases combining Lean Production and automation or Industry 4.0 are given in Sect. 3.4. Finally, conclusions are presented in Sect. 3.5.

3.2 Literature Review

Some research has been made to emphasize the interaction between Lean Manufacturing and Industry 4.0. For example, Sanders et al. (2016) analyzed the link between Industry 4.0 and lean manufacturing and investigates whether Industry 4.0 is capable of implementing lean. A methodology has been proposed to integrate Lean Manufacturing and Industry 4.0 with respect to supplier, customer, process as well as human and control factors. The authors also stated that researches and publications in the field of Industry 4.0 held answers to overcome the barriers of implementation of lean manufacturing. Similarly, Rüttimann and Stöckli (2016) discussed how Lean Manufacturing has to be regarded in the context of the Industry 4.0 initiative. Sibatrova and Vishnevskiy (2016) suggested an integrated Lean Management and Foresight, which considers the conditions of trends in Industry 4.0, human and time resources. Doh et al. (2016) not only reviewed the relevant literature of the industrial revolution to the new industry 4.0, but also added the needs of automation use in lean production systems and supply chain characterization with the aim of developing a framework for the integration of information systems and technologies. Blöchl and Schneider (2016) devised a new simulation game with the learning focus on Lean Logistics with Industry 4.0 components to teach the adequate application of Industry 4.0 technology in production logistics. Veza et al. (2016) made analysis of global and local enterprises which is based on literature review and questionnaires in order to develop Croatian model of Innovative Smart Enterprise (HR-ISE model). In the study, the selection of six basic Lean tools is made, and foundations of generic configuration of HR-ISE model are defined. Rauch et al. (2016) presented an axiomatic design oriented methodology which can be regarded as a set of guidelines for the design of Lean Product Development Processes. Linked with Industry 4.0, these guidelines show how a lean and smart product development process can be achieved by the use of advanced and modern technologies and instruments. Similarly, Synnes and Welo (2016) discussed organizational capabilities and tools required to enable transformation into Industry 4.0 through Integrated Product and Process Design. Biedermann et al. (2016) stated that maintenance needs to change to meet the requirements of Industry 4.0 and emphasized the necessity of knowledge and data management for improving predictive maintenance performance. Dez et al. (2015) proposed a novel Lean shop floor management system, namely Hoshin Kanri Tree (HKT). The authors also noted that the standardization of communication patterns by HKT technology should bring significant benefits in value stream performance, speed of standardization and learning rates to the Industry 4.0 generation of organizations.

What is more, it will be meaningful to mention about studies which addressed the interaction between automation and lean manufacturing. Automation cannot be excluded from lean principles since it has a significant role in lean manufacturing in terms of enhancing quality (Gingerich 2017; Weber 2016), reducing physical stress among shop-floor workers (Burg 2017), and generating flow (Harris and Harris 2008). Lean automation has been applied and evaluated for manufacturing (Bilberg 2006; Danielsson et al. 2010; Seifermann et al. 2014) and demanufacturing cells (Kuren 2003). On the other hand, Coffey and Thornley (2006) analyzed the reason of giving priority to managing the manual component instead of automation in car assembly in the subsequent years. In recent studies, Kolberg and Zühlke (2015) provided an overview over existing combinations of Lean Production and automation technology. Furthermore, the authors focused on major Industry 4.0 corner stones and links them to the well-proven Lean approach. Kolberg et al. (2016) described the ongoing work towards a common, unified communication interface to digitise Lean Production methods. Based on the Model-View-Controller-pattern, an architecture for the Cyber Physical Systems to loosely couple workstations to vendor-independent third-party solutions has been introduced. Gingerich (2017) presented a helpful approach to choose the right Lean production technologies by assessing the flow in terms of production volume and product mix and charting these factors against different types of production solutions. In this study, the matrix that Bosch Rexroth has created plots the four broad types of most commonly used production systems, namely manual production systems, modular production systems with some automation, standalone automated production cells, and fully automated production lines. Andrecioli and Lin (2008) presented the influence of cycle time variation and number of kanbans on the throughput of manufacturing cell supported by simulation tools so as to attain factory automation and introduced the concept of "Moving Bottleneck". From lean logistics viewpoint, Kamaryt et al. (2014) dealt with the comparison of the two logistic automation technologies, namely the AirMove system and Automatic Guided Vehicles, in crucial aspects of lean management quality-cost-time.

3.3 The Proposed Methodology

First, a brief information about the basic concepts of lean philosophy and lean production systems will be presented. Lean philosophy primarily aims at elimination of all activities that consume time and resources but do not add value to the physical completion of the products (Womack and Jones 2010). These activities are called waste or in Japanese muda, and termed as non-value adding activities. Here, the value is defined from the end customers' point of view and product specific. Hence, a value-adding activity is the one that contributes to the physical completion of the product and the one that the customer may want to pay for (Womack and Jones 2010). According to the lean philosophy, the wastes are intended to be eliminated. However, sometimes some of the wastes seem to be inevitable with the current technologies or manufacturing assets (Womack and Jones 2010). For instance, while switching from one product to another, a setup time can be unavoidable. Besides, there are other wastes that can be immediately eliminated by implementing lean tools and techniques.

According to the lean philosophy, there are seven traditional wastes or non-value adding activities common within the manufacturing systems. These are over-production, transportation, motion, waiting, inventory, unnecessary processing and defective parts/products (Ohno 1988). Later, Womack and Jones proposed that the products or services that do not meet the customer expectations should be regarded as a kind of waste (Womack and Jones 2010). Overproduction waste includes producing items for which there is no order or requirement (Liker 2004). This is the worst kind of waste, since it causes other wastes to occur. Due to the overproduction, large amount of inventory accumulates, excess amount of staff is employed; excess storages spaces are occupied, and so on. Inventory waste is linked to the overproduction, and also includes excess raw material, work-in process and finished goods inventory holding. Besides, excess inventory hides the problem within the production system such as frequent machine breakdowns, long setup times, defective parts.

Waste of waiting includes workers watching the running machines or waiting for the required materials, vehicles, tooling; and parts waiting for the rest of the batch to be completed. These frequently occur due to line imbalance, bottleneck resources, large batch production or late delivery of the materials by the suppliers. Transportation wastes include material handling for long or excessive distances.

These occur due to poor facility layout planning, and use of central warehouses rather than decentralized mini storages (Satoglu et al. 2006). Process wastes refer to performing additional steps due to poor tool design, product design or process design (Liker 2004). Wastes of Defectives imply unqualified parts or products, where some of them are scrap and others can be corrected by rework, but these defectives cause use of additional resources to either remanufacture the part or to make a rework. Besides, Motion wastes include unnecessary movement during the work, such as movements while looking for, reaching for, or stacking the tooling and parts.

There are several lean tools and techniques that can be utilized for waste elimination. The lean tools/techniques and the wastes that help eliminating are shown in Table 3.2.

On the other hand, there exist various advanced Industry 4.0 technologies and cyber-physical systems that can be employed for waste elimination in advanced manufacturing systems. The most fundamental technologies and the associated waste types that these technologies help for reduction are depicted in Table 3.3. Additive manufacturing (3-D Printing) may facilitate the design and manufacturing of the products customized according to the customer expectations and as much as needed, in suitable design complexity, customization and production volume conditions (Conner et al. 2014). Hence, it may reduce the lead time of the products and help in relieving waiting, inventory, unnecessary processing, overproduction wastes and defective parts/products.

Besides, the Augmented Reality technology in which the 3D virtual objects are integrated into a real environment in real time (Azuma 1997) can be employed for visualization of the operational work-instructions and specifications of qualified parts/products. It may help for eliminating the wastes of motion and transportation wastes and defectives.

In addition, the Digital Twin paradigm started to be mentioned frequently (Tuegel et al. 2011) that shows the potential of the Simulation and Virtualization. The simulation model of the factories and manufacturing processes can be constructed in virtual environment for quantifying and observing the alternative system designs. Cloud computing capabilities can be even enhanced for distant Modeling and Simulation of the manufacturing systems (Calheiros et al. 2009). By means of Simulation and Virtualization, the transportation, waiting, processing and defective product wastes of the manufacturing systems can be reduced.

Besides, Adaptive Robotics is an important aspect of the Smart Factories, such that the robots can adapt to the dynamic environmental conditions or parameters of the manufacturing system (Mahdavi and Bentley 2006). Adaptive Robotics enhances elimination of several types of wastes, including transportation, motion, waiting, processing and defectives.

Internet-of the Things (IoT) enhances integration among the physical systems within a manufacturing system by means of sensors and wireless communication technologies (Gubbi et al. 2013). Hence, it and may reduce the wastes of waiting, unnecessary inventory, overproduction and the defectives.

Table 3.2 The Seven Wastes versus Lean Tools/Techniques

	Cellular manufacturing	Setup reduction	Quality control	TPM	Production smoothing	Kanban	WIP reduction	Supplier development	Jidoka	CIM
Overproduction	✓	✓				✓	✓	✓		
Transportation	✓									
Motion	✓								✓	✓
Waiting	✓	✓		✓	✓	✓		✓	✓	✓
Inventory	✓	✓	✓		✓	✓	✓	✓		
Unnecessary processing		✓								✓
Defectives		✓	✓	✓				✓	✓	✓

Table 3.3 Seven wastes and advanced Industry 4.0 technologies

	Additive manufacturing (3-D printing)	Augmented reality	Simulation & virtualization	Adaptive robotics	IoT	Data analytics	Cloud computing
Transportation		✓	✓	✓		✓	
Motion	✓	✓		✓			✓
Waiting	✓		✓	✓	✓	✓	✓
Inventory	✓				✓	✓	
Unnecessary processing	✓		✓	✓			✓
Overproduction	✓				✓	✓	
Defectives	✓	✓	✓	✓	✓	✓	

When Data Analytics is applied to the manufacturing systems, it is expected to facilitate prediction of failures. It is also called Predictive Analytics (Lee et al. 2013). Hence, by means of preventing failures wastes of waiting, excess inventory (or safety stock), over-production, defectives and transportation costs can be reduced.

Cloud computing is an emerging parallel and distributed system that consists of interconnected and virtualized computers employed based on service-level agreements between the service provider and the customers (Buyya et al. 2008). It provides significant benefits to the companies by freeing them from setting up basic hardware and software infrastructures and the associated investments (Calheiros et al. 2009). Tao et al. (2014) pointed out that Cloud computing and IoT, together, enhances intelligent perception and communication on an M2M basis and on-demand and effective use of the resources, in the manufacturing systems. By means of efficient use of manufacturing equipment, wastes of waiting, unnecessary processing and motion can be eliminated to a certain extent.

After discussing the possible benefits of waste elimination provided by the Industry 4.0 technologies, the means of implementing these technologies integrated with the Lean tools is discussed. Figure 3.1 illustrates this projected relationship between the Lean tools/techniques and the advanced technologies. The figure is like a ladder implying that the lean tools/techniques should be implemented in a sequential manner. First, the layout of the manufacturing system should be converted into a Cellular manufacturing system that aims production of the product families by the autonomous and dedicated cells that are equipped with all required resources (Durmusoglu and Satoglu 2011). While implementing the Cellular

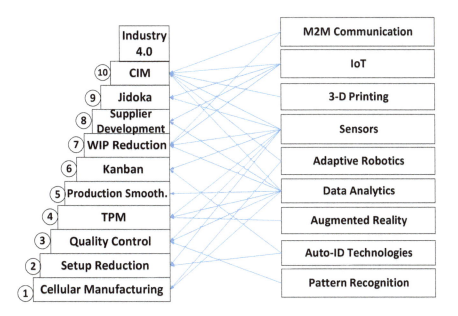

Fig. 3.1 Lean tools and Industry 4.0 technologies ladder

Manufacturing system, Data analytics can be utilized by performing a clustering analysis on the parts-machine matrices, to construct the manufacturing cells. Besides, Adaptive Robotics can be employed for an enhanced material handling, and parts loading-unloading.

For the setup reduction purposes, sensors that detect the components of the machines such as the dies, blades etc. can speed up the internal setup operations and protect the operators from accidents. Hence, occupational safety will be improved. Besides, Adaptive Robotics can also be implemented for Setup Reduction.

Quality Control and fool-proof mechanisms (Poka Yoke) are other important aspects of lean production systems. To prevent the defective parts and products, patter recognition technology can be employed to visually detect the defective parts by cameras. Besides, the Augmented Reality can be employed for giving instructions to the operators about the operations visually to prevent them from ill-performing the tasks. Sensors can be employed as fool-proof mechanisms or devices (poka-yokes).

Total Productive Maintenance (TPM) intends to improve the overall equipment effectiveness of the machines that includes reduction of time, speed and quality losses (Ahuja and Khamba 2008). Time losses are mainly due to the machine setup and failures. The machine failures can be prevented by auto-maintenance and parts replacement improving the machines status and by slowing its degradation. During the maintenance activities, Augmented Reality can be utilized to guide the operators for the maintenance activities. Besides, sensors keeping track of vibration, noise and heat help the operators detect the abnormal conditions before failure.

Production Smoothing is a production scheduling activity where the same quantities of a part or product is tried to be produced on a daily or hourly basis, as far as possible. Data Analytics is a suitable tool for analyzing the demand frequency coming from the customers.

Kanban is a lean production tool where the pull-production control is performed. This means a new batch of parts/products are produced only when needed and as much as needed (Satoglu and Ucan 2015). Traditionally, a barcode is printed on a kanban. However, thanks to the advanced Auto-ID Technologies, instead of scanning of barcodes of many kanbans, RFID tags are detected by the readers and a quick communication between the stages can be achieved.

For the Work-in Process (WIP) Reduction among the machines, a better M2M Communication, IoT, sensors and data analytics should be utilized. As the M2M communication by means of sensors is performed, each machine sense the availability state of the following machine and decides to start or stop production not to exceed the buffer area capacity. This mechanism basically reduces the WIP quantities. Besides, by means of Data analytics, the cycle times and the failure characteristics of the machines can be analyzed and the buffer area capacities among the machines can be adjusted.

For the Supplier Development purposes, a better Data analytics should be employed for better analyzing the demand data. To reach a better coordination and communication among the supplier and customer parties, IoT technologies should be employed.

Jidoka means automation with human touch (Liker and Morgan 2006). In other words, the manufacturing system employs the automation technologies under the supervision of the workers. So, while implementing the Jidoka, sensors and IoT can be employed.

While converting system into a Computer-Integrated Manufacturing System, M2M Communication, sensors, IoT, 3-D Printing, adaptive Robotics and Data Analytics can be employed for achieving more benefit out of the advanced manufacturing technologies.

3.4   Automation Based Lean Production Applications

Industry 4.0 technologies and automation can be applied to several methods of Lean Production. The following section describes examples of possible combinations.

E-Kanban Systems: The digitalization of the Kanban system is known already since several years. Conventional, physical cards for an order-oriented production control are replaced by virtual Kanban (Lage Junior and Filho 2010). Depending on the implementation of this so called e-Kanban system, missing or empty bins are recognized automatically via sensors. The e-Kanban system sends a virtual Kanban to trigger replenishment. By using Information Communication Technology, lost Kanban do not cause mistakes in production control anymore as long as inventory in manufacturing execution system matches real inventory. In addition, adjustments of Kanban due to changes in batch sizes, processes or cycle times are easily possible. Thermo King, an international manufacturer of temperature control systems, has implemented active RFID for what it calls an "e-Kanban" parts replenishment system at its factory in Galway, Ireland. When available parts at assembly stations drop to pre-set levels, workers simply push a call button that is integrated to an active RFID tag. The tag signal is received by the facility's wireless LAN. The tag's transmission indicates the part number and location that require replenishment, and the system software orders a batch to be delivered to the appropriate workstation.

Automation in Error-Proofing: Magna T.E.A.M. Systems makes widespread use of bar coding technology to eliminate human error and production mistakes. Operators scan bar codes wrapped around their wrists to ensure that they're assembling the correct product on the high-mix production line. All operators use bar codes to log into their workstation so that we have a record of who's building what part. Electronic work instructions displayed at every workstation address error proofing and serve as a visual aid to operators. The paperless system is a way for team members to interact while building a part in Magna. Because of all the nuances involved in high-mix assembly, assemblers must constantly look at the screen to ensure that they are using the correct component while correctly building to standardized work and validating torque values via sensors. Process control operation and monitoring in the Magna plant is handled by the MES system. The entire sequencing process is automated, including everything from receiving the build-order broadcast to shipping the bill of lading. In addition, electric fastening tools, bar code scanners and RFID tags are tied into the system for error proofing (Weber 2016).

Chaku Chaku Lines: In 2012, University of South Denmark together with toy manufacturer Lego A/S developed approaches for integrating automation technology in u-shaped assembly stations, also known as Chaku Chaku lines. Especially human machine interaction was in the focus of this project. As a result, they developed a local order management system which shifts typical tasks of ERP systems to employees at Chaku Chaku lines. According to them, automation of value adding tasks is particularly very reasonable due to the fact that investments are amortized within shorter time. Besides, the repeat accuracy and precision of machines is higher than of humans. On the other hand, complex processes, exception handling and logistic tasks are typical functions where automation is not reasonable (Bilberg and Hadar 2012). Moreover, The ongoing research project Lean Intelligent Assembly Automation also addresses Chaku Chaku lines. The consortium, which consists out of e.g. Adam Opel AG and Fraunhofer Institute for Manufacturing Engineering and Automation IPA, develops robot-based solutions to support employees in assembly tasks within Chaku Chaku lines. Objective is to enrich manual assembly tasks in order to make them more lucrative for bigger batches (Kolberg and Zühlke 2015).

iBin System: In 2013, Würth Industrie Services GmbH & Co. KG presented the optical order system iBin as an extension for Kanban bins (Fig. 3.1). A camera in the module detects the charging level of the bin and iBin reports wireless the status to an inventory control system. Besides, iBin is also able to send orders automatically to suppliers. As a result, buffer stock can be reduced and spare parts can be scheduled order-oriented (Würth Industrie Service GmbH & Co. KG 2013).

QR code integrated Milkrun System: Wittenstein AG and BIBA—Bremer Institut für Produktion und Logistik GmbH work among others in the state-funded project CyProS on a flexible material supply system for production lines. Instead of fixed intervals, an IT system calculates round trip intervals for the transport system based on real-time demands. In the first prototype, collection of data during this so-called milk run is done by scanning QR codes. Interaction with employees of the transport system is realized by conventional tablet PCs. With this order-oriented material supply system stretches of way can be reduced by approximately 25% at the same level of supplier's reliability (Kolberg and Zühlke 2015).

Pick-by-Vision: In the DHL application, the warehouse worker sees the physical reality of the aisles and racks in front of him just as he could if he were not wearing a head-mounted display, but this is augmented by a superimposed AR code in the form of a graphical work instruction, which appears after he scans the bar code at the storage location with his smart glasses. This code tells him where to go, how many items to pick and even where to place in his trolley. With the pilot project complete, DHL is evaluating the operational suitability and economic feasibility of adopting augmented-reality vision picking. Meanwhile, its Trends Research team has already identified other logistics activities that could be enhanced by a judicious dose of AR technology (Url-1).

Order-oriented control system: A Smart Product could contain Kanban information to control production processes. An example of a completely de-central controlled production based on Smart Products was demonstrated by *SmartFactoryKL* at Hannover Messe 2014 in Germany. The presented working stations produced autonomously according to a work schedule on the product. Although it was push-controlled production, this concept could be adopted for an order-oriented control system (Kolberg and Zühlke 2015).

Augmented Reality based work standardization: Project MOON (asseMbly Oriented authOring augmeNted reality) is developed by AIRBUS Military. MOON uses 3D information from the industrial Digital Mock-Up to generate assembly instructions and their deployment by applying Augmented Reality technology. A prototype was developed for the electrical harness routing in the frame 36 of the AIRBUS A400M (Servan et al. 2012).

Plug'n'Produce workstations: Industry 4.0 could furthermore support Lean Production's requirement for a flexible, modular production. Since several years *SmartFactoryKL* demonstrates modular working stations based on standardized physical and IT interfaces, which can be flexibly reconfigured to new production lines via Plug'n'Produce. According to the Single-Minute-exchange-of-Die (SMED) principle, setup time should be reduced to less than ten minutes. Plug'n'Produce transfers SMED from a single working station to whole production lines (Kolberg et al. 2016).

Smart Planner: Traditional Kanban systems with fixed amount of Kanban, fixed cycle times and fixed round trips for transporting goods turn into dynamical productions automatically adopting to current production programs via this system. Decentralized, in working stations integrated Cyber Physical System could nego- tiate cycle times and thus find the optimum between highest possible capacity utilization per working station and a continuous flow of goods. Within the state-funded project RES-COM, *German Research Center for Artificial Intelligence (DFKI) GmbH,* already demonstrated how a semantic description of working stations supports optimization of production processes by different business objectives, like throughput time or efficacy (Kolberg and Zühlke 2015).

Automatic Mold-Change System: At K 2016 in Dusseldorf, Staubli of Germany (U.S. office in Duncan, S.C.) demonstrated complete hands-off mold changing in less than 2 min, and company spokesmen said the system can get that down to 1 min. A mold table on rails carried a preheated mold into position beside the press. A sensor in the cart read the mold setup parameters from a chip in the mold. For the mold already in the press, all power and data connections were disconnected automatically within 3 s. A single manual lock disconnected all water lines. Ejectors retracted automatically, and the Staubli magnetic plates declamped the mold in another 3 s. The mold then slid out of the machine onto the cart, which indexed forward so the new mold could slide into place, clamp on magnetically, and make all the utilities connections. The final step was automatic exchange of robot grippers for the new job. Staubli sources said this system is being used at a new Plastic Omnium plant in Mexico, where several rail-mounted mold tables exchange 60-ton molds in 3700-ton presses in under 2 min. (Url-2).

Digitized Heijunka: Besides the flexible material supply system, WITTENSTEIN AG digitised the Heijunka-Board. Heijunka, also known as levelling, describes a method for converting customer orders into smaller, recurring batches (Verein Deutscher Ingenieure e.V. 2013). This can be realised using boards holding Kanban cards to control the production programme. Instead of conventional boards, displays with graphical user interfaces (GUI) which are connected to the production line and MES have been set up. By this, information flows and efforts for updating the board can be shortened (Kolberg et al. 2016).

Predictive Maintenance: Condition monitoring, data analytics and early prediction of the failures increases the uptime and overall equipment effectiveness (Bal and Satoglu 2014). For this purpose, predictive maintenance practices in manufacturing facilities have increased. In an oil and gas industry where the equipment are in remote locations, the oil-fields have been digitized by means of sensors. The name of the software platform is MAPR Distribution Including Hadoop® (MAPR 2015). In this architecture, the data collected by the sensors are transmitted to the time-series database by means of a Web service. The data in the database is analyzed by means of predictive algorithms and when a critical level of measure is detected, the maintenance is scheduled. Besides, for any manufacturing plant, electrical motors are essential elements, and they are mainly affected by the bearing failures. The bearing failures can be detected in advance by means of sensors that keep track of the vibration trend as a time-series. Alert, warning and fault levels of vibration velocity are determined so as to early predict and prevent the failure of the components.

3.5 Conclusion

The approach used in this paper answers a significant part of this question, and illustrates that lean manufacturing and Industry 4.0 are not mutually exclusive; they can be seamlessly integrated with each other for successful production management. This paper analyses the researches and publications concerned in the field of Industry 4.0, and identifies how they act as supporting factors for implementation of lean manufacturing.

In the previous chapters the terms of Lean Automation and Industry 4.0 have been described. Selected examples showed that the integration of innovative automation technology in Lean Production is an up to date and promising topic.

Industry 4.0 will not solve the problems of mismanaged and weakly-organized manufacturing systems. Its tools should be applied to lean activities which are performed successfully before automatization. In addition, effective information flow should be maintained effective before introducing ICT. In this context keeping the data in a right and current manner is an important critical success factor in both Industry 4.0 and Lean Production.

It is possible to see many examples high level automation applications in production facilities that do not require this level automation. There is no doubt that the right level of automation has a place in lean production, but understanding the impact of the various forms of automation and machine design on a lean production system is imperative to creating a flexible, efficient, world-class production system. A lean production system should be designed to flow, and automation should be selected after deciding how best to improve flow and fit into the flow. It should be stated that after implementing lean improvements, selective automation has the potential of adding value and however, it reduces human variability.

In addition, erroneous and prejudiced habits and waste-accustomed behaviors of employees about working method is a critical problem to be addressed in the design of manufacturing systems. In this context, the strategy should be to change the process of thought of people after altering their behavior with the help of a business discipline which does not distress people. Therefore, Lean Management System including work standardization and visual control is suggested to achieve this strategy.

References

Ahuja IPS, Khamba JS (2008) Total productive maintenance: literature review and directions. Int J Qual Reliab Manage 25(7):709–756

Andrecioli R, Lin YJ (2008) Lean manufacturing and cycle time variability analysis for factory automation using simulation tools. Proceedings of IMECE2008, Boston, Massachusetts, USA, October 31–November 6

Azuma RT (1997) A survey of augmented reality. Presence: Teleoperators virtual Environ 6(4):355–385

Bal A, Satoglu SI (2014) Maintenance management of production systems with sensors and RFID: a case study. In: Global Conference on Engineering and Technology Management (GCETM), 82–89

Biedermann H, Kinz A, Bernerstätter R, Zellner T (2016) Lean smart maintenance—Implementation in the process industry. Prod Manage 21(2):41–43

Bilberg A, Hadar R (2012) Adaptable and reconfigurable LEAN automation—a competitive solution in the western industry. 22th international conference on Flexible Automation and Intelligent Manufacturing (FAIM), Helsinki, Finland

Bilberg A (2006) Flexible hybrid lean automation. Proceedings of the 17th International DAAAM Symposium, 35–36

Blöchl SJ, Schneider M (2016) Simulation game for intelligent production logistics—The PuLL® learning factory. Procedia CIRP 54:130–135

Burg J Lean principles don't exclude automation. Manuf Eng. http://www.sme.org/MEMagazine/Article.aspx?id=21336&taxid=1428. Accessed 5 Jan 2017

Buyya R, Yeo CS, Venugopal S (2008) Market- oriented cloud computing: vision, hype, and reality for delivering IT services as computing utilities. In: Proceedings of the 10th IEEE international conference on high performance computing and communications

Calheiros RN, Ranjan R, De Rose CA, Buyya R (2009) Cloudsim: a novel framework for modeling and simulation of cloud computing infrastructures and services. arXiv prepr arXiv:0903.2525

Coffey D, Thornley C (2006) Automation, motivation and lean production reconsidered. Assembly Autom 26(2):98–103

Conner BP, Manogharan GP, Martof AN, Rodomsky LM, Rodomsky CM, Jordan DC, Limperos JW (2014) Making sense of 3-D printing: creating a map of additive manufacturing products and services. Add Manuf 1:64–76

Danielsson F, Svensson B, Gustavsson S (2010) A flexible lean automation concept for robotized manufacturing industry. Proceedings of the 11th Middle Eastern Simulation Multiconference, 101–104

Dez JV, Ordieres-Mere J, Nuber G (2015) The hoshin kanri tree. Cross Plant Lean Shop Floor Manag, Procedia CIRP 32:150–155

Doh SW, Deschamps F, Pinhero De Lima E (2016) Systems integration in the lean manufacturing systems value chain to meet industry 4.0 requirements. In: Borsato M. et al. (eds) Transdisciplinary engineering: crossing boundaries, 642–650

Durmusoglu MB, Satoglu SI (2011) Axiomatic design of hybrid manufacturing systems in erratic demand conditions. Int J Prod Res 49(17):5231–5261

Gingerich K (2017) Lean production and automation. http://www.plantengineering.com. Accessed 03 Jan 2017

Gubbi J, Buyya R, Marusic S, Palaniswami M (2013) Internet of Things (IoT): a vision, architectural elements, and future directions. Future gener comput syst 29(7):1645–1660

Harris R, Harris C (2008) Can automation be a lean tool? Manuf Eng. http://www.sme.org. Accessed 03 Jan 2017

Kamaryt T, Kostelný V, Hurzig A, Müller E (2014) Using innovative transportation technologies and automation concepts to improve key criteria of lean logistics.In: Proceedings of the 24th international conference on flexible automation and intelligent manufacturing: capturing competitive advantage via advanced manufacturing and enterprise transformation, 377–384

Kolberg D, Knobloch J, Zühlke D (2016) Towards a lean automation interface for workstations. Int J Prod Res. doi:10.1080/00207543.2016.1223384

Kolberg D, Zühlke D (2015) Lean Automation enabled by Industry 4.0 Technologies, IFAC-Papers. Online 48(3):1870–1875

Kuren MBV (2003) A lean framework for prototyping demanufacturing work cell automation. Proceedings in IEEE/ASME international conference on Advanced Intelligent Mechatronics, 663–668

Lage Junior M, Filho GM (2010) Variations of the Kanban system: literature review and classification. Int J Prod Econ 125:13–21

Lee J, Lapira E, Bagheri B, Kao HA (2013) Recent advances and trends in predictive manufacturing systems in big data environment. Manuf Lett 1(1):38–41

Liker JK (2004) The toyota way-14 management principles from the World's greatest manufacturer. McGraw Hill

Liker JK, Morgan JM (2006) The Toyota way in services: the case of lean product development. The Acad of Manag Perspect 20(2):5–20

Mahdavi SH, Bentley PJ (2006) Innately adaptive robotics through embodied evolution. Auton Robots 20(2):149–163

MAPR (2015) Predictive maintenance using hadoop for the oil and gas industry. https://www.mapr.com/sites/default/files/mapr_whitepaper_predictive_mainnance_oil_gas_051515.pdf. Accessed 24 Feb 2017

Ohno T (1988) Toyota production system: beyond large-scale production. CRC Press

Rauch E, Dallasega P, Matt DT (2016) The way from Lean Product Development (LPD) to Smart Product Development (SPD). Procedia CIRP 50:26–31

Rother M, Harris R (2001) Creating continuous flow: an action guide for managers, engineers & production associates. Lean Enterprise Institute

Rüttimann BG, Stöckli MT (2016) Lean and Industry 4.0-twins, partners, or contenders? a due clarification regarding the supposed clash of two production systems. J Serv Sci Manag 9: 485–500

Sanders A, Elangeswaran C, Wulfsberg J (2016) Industry 4.0 implies lean manufacturing: research activities in Industry 4.0 function as enablers for lean manufacturing. J Ind Eng Manag 9(3):811–833

Satoglu SI, Durmusoglu MB, Dogan I (2006) Evaluation of the conversion from central storage to decentralized storages in cellular manufacturing environments using activity-based costing. Int J Prod Econ 103(2):616–632

Satoglu SI, Ucan K (2015) Redesigning the material supply system of the automotive suppliers based on lean principles and an application. Proceedings of international conference on Industrial Engineering and Operations Management (IEOM), 1–6

Seifermann S, Böllhoff J, Metternich J, Bellaghnach A (2014) Evaluation of work measurement concepts for a cellular manufacturing reference line to enable low cost automation for lean machining. Procedia CIRP 17:588–593

Serván J, Mas F, Menéndez JL, Ríos J (2012) Assembly work instruction deployment using augmented reality. Key Eng Mater 502:25–30

Sibatrova SV, Vishnevskiy KO (2016) Present and future of the production: integrating lean management into corporate foresight, Working Paper, National Research University Higher School of Economics, WP BRP 66/STI/2016

Synnes EL, Welo T (2016) enhancing integrative capabilities through lean product and process development. Procedia CIRP 54:221–226

Takeda H (2006) The synchronized production system: going beyond just-in-time through kaizen. KoganPage, London

Tao F, Cheng Y, Da Xu L, Zhang L, Li BH (2014) CCIoT-CMfg: cloud computing and internet of things-based cloud manufacturing service system. IEEE Transactions on Industrial Informatics, 10(2):1435–1442

Tuegel EJ, Ingraffea AR, Eason TG, Spottswood SM (2011) Reengineering aircraft structural life prediction using a digital twin. Int J Aerosp Eng. http://dx.doi.org/10.1155/2011/154798

Url-1, https://logisticsviewpoints.com/2015/04/16/picking-with-vision/

Url-2, http://www.ptonline.com/articles/fully-automatic-mold-change-in-under-2-min

Url-3, http://www.rfidjournal.com/articles/view?7123

Veza I, Mladineo M, Gjeldum N (2016) Selection of the basic lean tools for development of croatian model of innovative smart enterprise. Tehnički vjesn 23(5):1317–1324

Weber A (2016) Automation and lean help magna stay flexible. Assembly. http://www.assemblymag.com/articles/93268-automation-and-lean-help-magna-stay-flexible. Accessed 30 Dec 2016

Womack JP, Jones DT (2010) Lean thinking: banish waste and create wealth in your corporation. Simon and Schuster

Würth Industrie Service GmbH & Co. KG iBin(R) stocks in focus—the first intelligent bin (2013)

Piercy N, Rich N (2009) Lean transformation in the pure service environment: the case of the call service centre. Int J Oper Prod Manag 29(1):54–76

Zuehlke D (2010) SmartFactory-Towards a factory-of-things. Annu Rev Control 34:129–138

Chapter 4
Maturity and Readiness Model for Industry 4.0 Strategy

Kartal Yagiz Akdil, Alp Ustundag and Emre Cevikcan

Abstract Companies that transform their businesses and operations regarding to Industry 4.0 principles face complex processes and high budgets due to dependent technologies that effect process inputs and outputs. In addition, since Industry 4.0 transformation creates a change in a business manner and value proposition, it becomes highly important concept that requires support of top management for the projects and investments. Therefore, it requires a broad perspective on the company's strategy, organization, operations and products. So, the maturity model is suitable for companies planning to transform their businesses and operations for Industry 4.0. It is a very important technique for Industry 4.0 in terms of companies seeking for assessing their processes, products and organizations and understanding their maturity level. In this chapter, existing maturity models for Industry 4.0 transformation are reviewed and a new Industry 4.0 maturity model is proposed.

4.1 Introduction

In today's world, economic challenges driven by technological and societal developments force industrial enterprises improve their agility and responsiveness in order to gain ability to manage whole value-chain. Hence, enterprises require assistance of virtual and physical technologies which provide collaboration and rapid adaption for their businesses and operations (Ganzarain and Errasti 2016). Implementation of Industry 4.0 strategies require wide applications in companies

K.Y. Akdil (✉) · A. Ustundag · E. Cevikcan
Faculty of Management, Department of Industrial Engineering,
Istanbul Technical University, Macka, 34367 Istanbul, Turkey
e-mail: akdilyagiz@gmail.com

A. Ustundag
e-mail: ustundaga@itu.edu.tr

E. Cevikcan
e-mail: cevikcan@itu.edu.tr

© Springer International Publishing Switzerland 2018   61
A. Ustundag and E. Cevikcan, *Industry 4.0: Managing The Digital Transformation*,
Springer Series in Advanced Manufacturing, https://doi.org/10.1007/978-3-319-57870-5_4

since executives from several industries are uncertain about outcomes of the
Industry 4.0 projects and investment costs and they have lack of knowledge in the
concept of Industry 4.0. Maturity models provide large scale of knowledge about
companies' current state and a path to pursue for implementation of Industry 4.0
strategies.

Nikkhou et al. (2016) stated that maturity term refers to being in a perfect
condition; also, it is an evidence of an achievement and it provides a guidance to
correct or prevent problems. Mettler (2009) defined maturity as a development of a
specific ability or reaching to a targeted success from an initial to an anticipated
stage. Proença and Borbinha (2016) reported that maturity can be used as evaluation criteria and described as being complete, perfect or ready and also used for
progression from a basic stage to more advanced final stage.

According to Tarhan et al. (2016) a maturity model represents desired logical
path for processes in several business fields which include discrete levels of
maturity. In addition, maturity models are defined as a valuable technique in order
to assess processes or organization from different perspectives (Proença and
Borbinha 2016). Backlund et al. (2014) also noted that maturity model frameworks
are becoming extremely important to assess organizations. Nikkhou et al. (2016)
described maturity models as a tool that can be used to describe perfect progression
to wanted change utilizing a few progressive phases or levels. Maturity models
enable organizations to audit and benchmark regarding to assessment results, to
track progress towards to desired level and to evaluate elements of organizations
such as strengths, weaknesses and opportunities by sequencing maturity levels in an
order from basic to advanced stage: Initial, Managed, Defined, Quantitatively
Managed and Optimizing (Proença and Borbinha 2016). According to Schumacher
et al. (2016) maturity models are positioned as a tool for comparing current level of
an organization or process to desired level in terms of maturity by conceptualizing
and measuring. In the article, the difference between readiness and maturity models
is explained in such a way that readiness models clarify whether organization is
ready to start development process or not; however, maturity models target to
demonstrate which maturity level the organization is in. In brief, Duffy (2001)
concluded that maturity models help organizations to decide when and why they
need to take an action to progress; in addition, teach organizations which actions
should be considered in order to achieve advanced maturity level. According to
author, organization must need the information obtained from maturity models to
compare its current state to the best-practices in related business fields. Therefore, a
value of a maturity model is measured by its usability on analysis and positioning.

In study of IBM in 2015, it is stated that Industry 4.0 transformation has many
difficulties because of the inadequacy of current IT technologies, lack of knowledge
and high investment costs (Erol et al. 2016). Digital transformation of companies'
businesses and operations is driven by investments in information and telecommunication technologies and new machineries. In addition, the need of the integration in current technologies, new machines and automated work processes
restrain horizontal and vertical integration along the value chain (Erol et al. 2016).

According to industry-wide interviews, when implementing Industry 4.0 in practice, following problems come in view (Schumacher et al. 2016):

- Lack of strategic guidance and the problem of perception about highly complex Industry 4.0 concept.
- Uncertainty about outcomes of Industry 4.0 projects in the matter of benefits and costs.
- Failure of assessing Industry 4.0 capability of company.

With regard to third problem, maturity models and assessment of Industry 4.0 maturity become highly important, since a lot of companies seem to struggle to initialize Industry 4.0 transformation.

The purpose of this chapter is to explain "maturity models"; discuss the encountered problems when implementing Industry 4.0 strategies; to explain reasons of utilization of these models and their benefits to Industry 4.0 strategies. The rest of the chapter is organized as follows. Section 4.2 includes detailed explanation of four maturity models in the literature. Then in Sect. 4.3, the comparison chart of analyzed Industry 4.0 maturity and readiness models is presented. Section 4.4 is organized to propose Industry 4.0 maturity model and an application in retail sector. The conclusion and future work are presented in final section.

4.2 Existing Industry 4.0 Maturity and Readiness Models

In this section, we performed analysis of several Industry 4.0 maturity and readiness models and assessment surveys. From these works, we derived concepts relevant for the structure of our model. Models and assessments are given below:

- IMPULS—Industrie 4.0 Readiness (2015)
- Industry 4.0/Digital Operations Self-Assessment (2016)
- The Connected Enterprise Maturity Model (2016)
- Industry 4.0 Maturity Model (2016).

4.2.1 IMPULS—Industrie 4.0 Readiness (2015)

Lichtblau et al. and other project partners performed Industry 4.0 workshops and literature researches to propose an Industry 4.0 readiness model. This model contains six levels of Industry 4.0 readiness given below:

- Level 0: Outsider
- Level 1: Beginner
- Level 2: Intermediate
- Level 3: Experienced

- Level 4: Expert
- Level 5: Top performer.

In this study, existing readiness model redesigned with contributions from workshops and formed in six Industry 4.0 dimensions by adding two dimensions to previous model. These dimensions (Lichtblau et al. 2015) are "Strategy and organization", "Smart factory", "Smart operations", "Smart products", "Data-driven services" and "Employees". Dimensions and associated fields of Industry 4.0 related with this model are given in Table 4.1.

Proposed readiness model is used in order to measure companies' Industry 4.0 readiness levels from 0 to 5 containing minimum requirements that companies should possess. Questionnaire of this readiness model measures structural characteristics of companies, their Industry 4.0 knowledge, their motivations and obstacles through Industry 4.0 journey. Assessment survey has 24 questions in total for related dimensions and a few questions about industry, size of domestic workforce and annual revenue. In order to measure and define Industry 4.0 readiness, five point Likert scale is used.

Company profiles are grouped under three titles to summarize results better (Lichtblau et al. 2015) such as Newcomers (level 0 and 1), Learners (level 2) and Leaders (level 3 and up). Newcomers consist of companies that have never initialized any projects or have studied a few projects. Learners consists a group of companies which initialized first projects related to Industry 4.0. Leaders is a group that contains level 3, 4 or 5 companies which are way ahead of other companies about Industry 4.0 implementation.

Companies' readiness levels are determined based on lowest level of associated field in the dimensions. Industry 4.0 dimensions are weighted on 100-point-scale. "Strategy and organization" has 25 point overall, "Smart factories" has 14 points

Table 4.1 Dimensions and associated fields of Industry 4.0 (Lichtblau et al. 2015)

Dimensions	Associated fields
Strategy and organization	Strategy Investments Innovation management
Smart factory	Digital modelling Equipment infrastructure Data usage IT systems
Smart operations	Cloud usage IT security Autonomous processes Information sharing
Smart products	Data analytics in usage phase ICT add-on functionalities
Data-driven services	Share of data used Share of revenues Data-driven services
Employees	Skill acquisition Employee skill sets

overall, "Smart products" has 19 points overall, "Data-Driven services" has 14 points overall, "Smart operations" has 10 points overall and finally "Employees" has 18 points overall.

As a result of measurements, company profiles are identified and main hurdles in the dimensions are listed. In the final stage, action plans are created for companies to help them reach level 5 Industry 4.0 readiness.

4.2.2 Industry 4.0/Digital Operations Self-Assessment (2016)

PwC published a report entitled "Industry 4.0: Building the digital enterprise" to provide companies comprehensive perspective on Industry 4.0 by representing its own maturity model and "Blueprint for Digital Success" given in Table 4.2.

In the first step of "Blueprint for Digital Success", PwC provides companies a maturity model to assess their capabilities. This maturity model is formed in four stages and seven dimensions. Stages are determined as below:

- Digital novice
- Vertical integrator
- Horizontal collaborator
- Digital champion.

PwC assess companies' maturity levels with seven dimensions such as "Digital business models and customer access", "Digitisation of product and service offerings", "Digitisation and integration of vertical and horizontal value chains", "Data and Analytics as core capability", "Agile IT architecture", "Compliance, security, legal and tax", "Organisation, employees and digital culture".

PwC enables companies to assess their Industry 4.0 maturity and map their results by online self-assessment tool. In the final stage of assessment, PwC provides companies an action plan to make them successfully reach high level of Industry 4.0 maturity.

Online self-assessment tool (PwC, 2016) has 33 questions in total for related dimensions and a few questions about industry, region, country and annual revenue to classify companies. In questionnaire, five point Likert scale is used for each question and radar graphic is provided at the end of the assessment.

Table 4.2 Blueprint for digital success (Geissbauer et al. 2016)

Practical steps					
Step 1	Step 2	Step 3	Step 4	Step 5	Step 6
Map out your Industry 4.0 strategy	Create initial pilot projects	Define the capabilities you need	Become a virtuoso in data analytics	Transform into a digital enterprise	Actively plan an ecosystem approach

4.2.3 The Connected Enterprise Maturity Model (2016)

The Connected Enterprise Maturity Model was developed by Rockwell Automation in 2014 and this model contains five stages and four technology focused dimensions. Stages in this model are given below (Rockwell Automation 2016):

- Stage 1: Assessment
- Stage 2: Secure and upgraded network and controls
- Stage 3: Defined and organized working data capital
- Stage 4: Analytics
- Stage 5: Collaboration.

The Assessment Stage of the Connected Enterprise Maturity Model evaluates all facets of an organization's existing OT/IT (Operational Technologies/Information Technologies) network with four dimensions such as "Information infrastructure (hardware and software)", "Controls and devices (sensors, actuators, motor controls, switches, etc.) that feed and receive data", "Networks that move all of this information" and "Security policies (understanding, organization, enforcement)". It is stated that a major challenge during assessment stage is potential hesitations on investing time for questioning practices that they have relied upon for years.

In Stage 2, OT/IT organization is being formed to deliver secure, adaptable connectivity between plant-floor operations and enterprise business systems after an assessment stage. Long-term upgrades begin and gaps and weaknesses of current operations are identified. In large-scale companies, outdated control and networks create a challenge for transformation as well as hesitations from executives and engineers who feel that current systems remain viable.

Stage 3, which the improvements made with the current data is in progress with Stage 2. At this stage, it is defined how the collected data will be processed and how to obtain optimum outcome from these data. Organized team ensures that new workflows, charts and responsibilities are set so that they are not overwhelmed by the company data pool thanks to Working Data Capital.

In Stage 4, focal point shifts towards continuous development with data. Analytics utilizing Working Data Capital will assist to pinpoint the greatest needs for real-time information and ensure the continuity of standardized protocols triggered by the data. In addition, these analytics provide information transfer about asset management for leadership team. Challenges during Stage 4 will be use of lots of unnecessary data and distrust of analytics.

In Stage 5, main idea is to provide collaboration between company and environment with the help of analytics and data sharing.

4.2.4 Industry 4.0 Maturity Model (2016)

Schumacher et al. (2016) used nine dimensions and sixty-two maturity items in order to assess companies Industry 4.0 maturity levels. Nine dimensions and maturity items are given in Table 4.3. Maturity levels are examined under five levels. According to this model, level 1 companies have lack of attributes the supporting concepts of Industry 4.0 and level 5 companies can meet all requirements of Industry 4.0.

In this maturity model, assessment surveys are made by using five point Likert scale for each closed ended question. After survey results weighted points are calculated and maturity levels of companies are determined. To determine maturity level of a company, an equation (Eq. 4.1) is used. In this equation, "M" corresponds to "Maturity", "D" corresponds to "Dimension", "I" corresponds to "Item", "g" corresponds to "Weighting Factor" and "n" corresponds to "Number of Maturity Item" (Schumacher et al. 2016).

$$M_D = \frac{\sum_{i=1}^{n} M_{DIi} * g_{DIi}}{\sum_{i=1}^{n} g_{DIi}} \tag{4.1}$$

Table 4.3 Dimensions and maturity items of Industry 4.0 Maturity Model (Schumacher et al. 2016)

Dimensions	Exemplary maturity item
Strategy	Implementation I40 (Industry 4.0) roadmap, Available resources for realization, Adaption of business models, …
Leadership	Willingness of leaders, Management competences and methods, Existence of central coordination for I40, …
Customers	Utilization of customer data, Digitalization of sales/services, Costumer's Digital media competence, …
Products	Individualization of products, Digitalization of products, Product integration into other systems, …
Operations	Decentralization of processes, Modelling and simulation, Interdisciplinary, interdepartmental collaboration, …
Culture	Knowledge sharing, Open-innovation and cross company collaboration, value of ICT in company, …
People	ICT competences of employees, openness of employees to new technology, autonomy of employees, …
Governance	Labour regulations for I40, Suitability of technological standards, Protection of intellectual property, …
Technology	Existence of modern ICT, Utilization of mobile devices, Utilization of machine-to-machine communication, …

4.3 Comparison of Existing Industry 4.0 Maturity and Readiness Models

In this section, maturity/readiness levels, dimensions and industry scope are compared between existing Industry 4.0 maturity and readiness models. In order to provide easy understanding, comparison table is given in Table 4.4.

4.4 Proposed Industry 4.0 Maturity Model

In order to facilitate different analyses of Industry 4.0 maturity, the proposed model includes a total of 13 associated fields which are grouped into 3 dimensions. Table 4.5 provides an overview on the dimensions together associated fields, related Industry 4.0 to support understanding. An assessment criterion of the maturity model is based on Industry 4.0 principles and technologies given in Table 4.6 for each associated field.

Smart products can do computations, store data and be involved in an interaction with their environment as well as they can give information about their identity, properties, status and history (Schmidt et al. 2015). These features create a chance to obtain data from products and interpret data to offer services. **Smart Products**

Table 4.4 Comparison of existing industry 4.0 maturity and readiness models

Maturity/readiness model	Maturity/readiness levels and dimensions	Industry scope
IMPULS—Industrie 4.0 Readiness (2015)	5-staged readiness model with 6 dimensions and 18 criteria. Development steps for each stage are clearly defined. Main hurdles, obstacles and action plans are determined for each stage	Focused on manufacturing and engineering industry. Limited application area
Industry 4.0/Digital operations self-assessment (2016)	4-staged maturity model with 7 dimensions. Undefined maturity criteria. Comments and short action plans are provided as a result of online assessment	Industry wide maturity model. Wide application area
The connected enterprise maturity model (2016)	5-staged maturity models with 5 dimensions and technology focus. Undefined maturity criteria. Lack of assessment tool	Focused on IT capability of companies. Lack of organization and operations dimension. Limited application area
Industry 4.0 Maturity Model (2016)	5-staged maturity model with 9 dimensions and 62 maturity items. Basic formulation for assessment	Focused on manufacturing industry. Comprehensive maturity model and assessment. Limited application area

Table 4.5 Proposed industry 4.0 maturity model

Dimensions	Sub-dimension	Associated fields
Smart products and services		Smart products and services
Smart business processes	Smart production and operations	Production, logistics and procurement
		R&D—Product development
	Smart marketing and Sales operations	After sales service
		Pricing/Promotion
		Sales and Distribution channels
	Supportive operations	Human resources
		Information technologies
		Smart finance
Strategy and Organization		Business models
		Strategic partnerships
		Technology investments
		Organizational structure and leadership

and Services dimension is formed to measure these features of companies' products and their service offerings driven by product data.

Smart Business Processes is formed as a dimension containing functional operations of companies to assess their maturity level regarding to Industry 4.0 principles and triggering technologies.

Strategy and Organization can be defined as an "input" for Industry 4.0 transformation where it is important to shape business and organization. Development of new smart products, data-driven services and smart business operations depend on generating suitable business models or transforming current one for Industry 4.0, investments in triggering technologies, collaboration with strategic partners which provides fast progression and organizational structure and leadership.

To identify Industry 4.0 maturity level of a company, four stages are used and answers of the assessment survey is evaluated regarding to these stages such as "Absence", "Existence", "Survival" and "Maturity". Each associated field's questions weighted between 0—"Absence" and 3—"Maturity" to determine a maturity level.

Level 0: Absence identifies a level of a company that does not meet any of the requirements for Industry 4.0. Some of the requirements are at low level.

Level 1: Existence is a maturity level where company has some pilot initiatives in its functional departments. Company provides products, but these products are not capable of being fully smart. Integration and automation levels are low and data collection/use levels are not enough to realize Industry 4.0 transformation. Digital technologies and cloud has not been implemented to all operations. Equipment infrastructure readiness is also at low level. Top management is considering implementing Industry 4.0 strategy with investments in a few areas. There are pilot initiatives to generate business models or transform current one. Organizational structure is not suitable enough.

Table 4.6 Industry 4.0 principles and technologies

Principles	Technologies
Real time data management (Collection/Processing/Analysis/Inference)	Adaptive robotics
	Data analytics and Artificial intelligence
Interoperability	Simulation
Virtualization	Embedded systems
Decentralized	Communication and Networking
Agility	Cybersecurity
Service oriented	Cloud
Integrated business processes	Additive manufacturing
	Virtualization technologies
	Sensors and Actuators
	RFID and RTLS technologies
	Mobile technologies

Level 2: Survival is a maturity level where company's products are capable of real time data management and being tracked through different sites; in addition, data-driven service offerings are at medium level. Company's business processes at medium level in terms of integration, data sharing/collection/use and agility. Processes are ready for decentralization and interoperability principle is implemented a few areas in company with support of digital technologies. Leadership is developing plans for Industry 4.0 and has made investments in a few areas. Company is considering new business opportunities at medium level and creating partnerships with other companies or academics. Organizational structure is suitable for initial Industry 4.0 projects and new business models are being generated.

Level 3: Maturity is a maturity level where company's products are defined as smart and data-driven services are provided high level. Company's business processes at high level in terms of integration, data sharing/collection/use and agility. Nearly all processes are capable of being decentralized and interoperability principle is implemented lots of areas in company with support of advanced digital technologies. Leadership team provides widespread support for Industry 4.0 and has made investments for nearly all departments. Organizational structure is suitable for managing transformation across the company. Company is creating lots of partnerships with companies, academics, suppliers and technology providers. Digital business models are integrated to company's current business models and company is generating revenue from these models.

Each associated field in this maturity model is graded with related survey questions by 0–3 points. After all, calculated points of associated fields are grouped under dimensions and sub-dimensions in order to identify maturity levels individually and overall. Equations to calculate maturity levels are given in Eqs. 4.1–4.3.

M   Maturity
D   Dimension
A   Associated Field
Q   Question Number
O   Overall

n   Number of Total Questions
m   Number of Associated Fields

$$M_{DAi} = \frac{\sum_{j=1}^{n} Q_{Aij}}{n} \tag{4.2}$$

$$M_D = \frac{\sum_{i=1}^{m} M_{DAi}}{m} \tag{4.3}$$

$$M_O = \min(M_1, M_2, M_3) \tag{4.4}$$

Table 4.7 Limit values to determine maturity level

Maturity level	Limit values	
	Low	High
Level 0: Absence	0.00	0.90
Level 1: Existence	0.90	1.80
Level 2: Survival	1.80	2.70
Level 3: Maturity	2.70	3.00

Table 4.8 Smart products and services maturity level requirements

Maturity level	Smart products and services
Level 0: Absence	Company does not meet any of the requirements for Industry 4.0. Some of the requirements are at low level
Level 1: Existence	Company's products are capable of communicating with other products/platforms, machines and external systems as well as collecting data Products can be tracked as they move between manufacturing and internal internal distribution sites Company offer service/insights for only its business according to data obtained from the product
Level 2: Survival	Company's products are capable of communicating and collecting data. In addition, products can keep data they collect on their system or in the cloud Product can perform descriptive, diagnostics and predictive data analysis. Products can be tracked as they move between manufacturing and distribution until they reach the customers DC Company offer service/insights for its business and customers according to data obtained from the product
Level 3: Maturity	Company's products are capable of communicating with other systems, collecting data and keeping it on their system or in the cloud. In addition, products have a platform on which the product or cloud applications are working Product can perform descriptive, diagnostics, predictive and prescriptive data analysis Products can be tracked along their complete lifecycle Company offer service/insights for its business, customers and partners according to data obtained from the product improvements

Table 4.9 Smart business processes maturity level requirements

Maturity level	Smart business processes
Level 0: Absence	Company does not meet any of the requirements for Industry 4.0. Some of the requirements are at low level
Level 1: Existence	Supply chain processes are integrated between company, suppliers and customers in terms of basic data sharing and communication There are a few software systems in use and production systems are partially automated at machine level. Operation process traceability is provided at machine level (partial) in the digital environment and end-to-end visibility is at low level as well as the production customization level. Data usage in new product development is at low level. Manufacturability and terms of use of the product is simulated during product development at low level In after sales services process, company benefits from few data and offer services in a few area. Triggering technologies (i.e. mobile and virtualization technologies, cloud) are not in use. A few analytics studies are conducted and data obtained from environment is not used in product pricing and dynamic pricing. Campaign systems and sales channels has low level integration and data analytics tools are not in use to measure campaign performances. Integration of communication channels and collaboration with partners are at low level. In Human Resources operations data is used in a few areas, but company does not share real-time data with field workers and e-learning is not an option. IT security solutions are planned or in progress for data through cloud services. IT dashboards are not in use and machines/systems can be controlled through IT to some extent. Automation of financial services is at low level and analysis generally basis on historical data. Triggering technologies (i.e. 3D printers, cloud, mobile and virtual technologies, etc.) are being used at low level
Level 2: Survival	Supply chain processes are integrated between company, and key strategic suppliers/customers in terms of data transfer. There are some software systems in use and production systems are exactly automated at machine level or partially automated at production line/cell level. Operation process traceability is provided at production line/cell level in the digital environment and end-to-end visibility is at medium level as well as the production customization level. Data usage in new product development is at medium level. Manufacturability and terms of use of the product is simulated during product development at medium level In after sales services process, company benefits from some data and offer services in areas. Triggering technologies (i.e. mobile and virtualization technologies, cloud) are in use. Analytics studies are conducted and data obtained from environment is used in product pricing and dynamic pricing. Campaign systems and sales channels has medium level integration and data analytics tools are in use to measure campaign performances. Integration of communication channels and collaboration with partners are at medium level. In Human Resources operations data is used in some areas, company shares real-time data with field workers and e-learning is an option. IT security solutions are in progress or implemented in terms of communications for in-house data exchange. IT dashboards are in use and machine to machine communications are available. Automation of financial services is at medium level and analysis generally basis on historical data. Triggering technologies (i.e. 3D printers, cloud, mobile and virtual technologies, etc.) are being used at medium level

(continued)

Table 4.9 (continued)

Maturity level	Smart business processes
Level 3: Maturity	Supply chain systems are fully integrated between company, suppliers and customers which provides real-time planning. There are lots of software systems in use and production systems are exactly automated at production line/cell level or partially automated in factory level. Operation process traceability is provided at factory level in the digital environment and end-to-end visibility is at high level as well as the production customization level. Data usage in new product development is at high level. Manufacturability and terms of use of the product is simulated during product development at high level.
	In after sales services process, company benefits from lots of data and offer services in wide range. Triggering technologies (i.e. mobile and virtualization technologies, cloud) are in use. Analytics studies are conducted and data obtained from environment is used in product pricing and dynamic pricing. Campaign systems and sales channels has high level integration and data analytics tools are in use to measure campaign performances. Integration of communication channels and collaboration with partners at high level. In Human Resources operations data is used in lots of areas, company shares real-time data with field workers and e-learning is an option. IT security solutions are implemented for data exchange with business partners. IT dashboards are in use and interoperability principle is applied completely. Automation of financial services is at high level and analysis generally basis on real-time data. Triggering technologies are being used at high level

Table 4.10 Strategy and organization maturity levels requirements

Maturity level	Strategy and organization
Level 0: Absence	Company does not meet any of the requirements for Industry 4.0. Some of the requirements are at low level
Level 1: Existence	Existing products and services are not compatible with digital business models which is supported with resources at low level
	There is an awareness of "as a service" business model and revenue is generated from data-driven services (0–2.5%)
	Company has partnerships with a few stakeholders and launched pilot initiatives. Leadership team is investigating potential benefits
	Company allocated low level of budget to technologies and is planning investments in a few functional operations; but cost/benefit analysis are not conducted
	Organizational structure is not suitable for transformation
	Only technology focused areas has employees with digital skills that are not allocated to specific Industry 4.0 projects
	Central IT departments are existed in the company where there is not any working environment where OT/IT units work together
	There is limited interaction between departments

(continued)

Table 4.10 (continued)

Maturity level	Strategy and organization
Level 2: Survival	Existing products and services are compatible with digital business models which is supported with resources at medium level
	There is a high awareness of "as a service" business model and revenue is generated from data-driven services (2.5–10%)
	Company has partnerships with some of stakeholders and strategy is formulated. Leadership team recognizes financial benefits and plans to invest
	Company allocated medium level of budget to technologies and investments are done in some of functional operations; in addition annual cost/benefit analysis are conducted
	Organizational structure is suitable for initial projects
	In most areas of the business have well developed digital skills that are allocated to specific Industry 4.0 projects in different units
	Local IT departments are existed in each area where there is a working environment where OT/IT units work together
	Departments are open to cross-company collaboration
Level 3: Maturity	Existing products and services are compatible with digital business models which is supported with resources at high level
	"As a service" has been implemented and is being offered. Revenue is generated from data-driven services (over 10%)
	Company has partnerships with lots of stakeholders and strategy is implemented or in implementation. Widespread support for the Industry 4.0 within leadership and across the wider business
	Company allocated high level of budget to technologies and invested in nearly all functional operations; in addition quarterly cost/benefit analysis are conducted
	Organization is well structured for transformation
	All across the business, cutting edge digital and analytical skills are prevalent and allocated to specific Industry 4.0 projects in same units
	IT experts attached to each department where there is a working environment where OT/IT units work together
	Departments are open to cross-company collaboration to drive improvements

To determine overall maturity level, limit values for each level is given in Table 4.7. Maturity levels of "smart products and services", "smart business" and "strategy and organization" are explained in Table 4.8, 4.9 and 4.10 respectively.

4.5 An Application in Retail Sector

The study was conducted in a retail company operating in Turkey. In this study, the questionnaires in Appendix were answered and according to the answers given, scores related to the relevant fields and dimensions were calculated according Eqs. 4.2–4.4.

Table 4.11 Smart products and services maturity score

| # of question | Score |
	Smart products and services
1	3.00
2	3.00
3	1.00
4	3.00
Maturity Score	2.50

Table 4.12 Smart business processes maturity score

| # of question | Score | | | | | | | |
	Production, Logistics and procurement	R&D— product development	After sales services	Pricing/ promotion	Sales and distribution channels	Human resources	Information technology	Smart finance
1	1.31	N/A	0.66	3.00	3.00	1.50	3.00	0.00
2	3.00	N/A	N/A	0.00	0.00	3.00	2.00	3.00
3	0.00	N/A	1.00	3.00	3.00	3.00	3.00	3.00
4	1.50	N/A		3.00	3.00		1.50	2.00
5	1.13	N/A		3.00	1.00			2.00
6	1.00				0.00			
7	0.60				0.00			
8	1.00							
9	0.00							
10	0.00							
Associated field score	0.95	N/A	0.83	2.40	1.43	2.50	2.38	2.00
Maturity score	1.60							

The scores corresponding to the answers given to the questions are shown on the Tables 4.11, 4.12 and 4.13. Since the firm operates in the retail sector, it has decided not to answer the questions about the field of "R&D—Product Development". So this field did not participate in the scoring account according to Eqs. 4.2–4.4.

According to an equation (Eq. 4.4), overall maturity level of a company is determined by minimum maturity level of dimensions. As we can see in Table 4.11, Table 4.12 and Table 4.13, minimum maturity level score is 1.14 which is calculated for Strategy and Organization dimension. Therefore, a retail company is at "Level 1: Existence" regarding to Industry 4.0 maturity. Summary of maturity scores is given in Table 4.14. Radar chart is provided in Fig. 4.1.

Table 4.13 Strategy and organization maturity score

	Score			
# of question	Business models	Strategic partnerships	Technology investments	Organizational structure and leadership
1	3.00	1.50	1.75	3.00
2	0.00	1.00	1.00	3.00
3	2.00	0.00	0.00	3.00
4	0.00		0.94	1.00
5	0.00			0.00
6				2.00
7				2.00
8				1.00
9				0.00
10				3.00
Associated field score	1.00	0.83	0.92	1.80
Maturity score	1.14			

Table 4.14 Maturity score and level table for each dimension/sub-dimension of a company

Dimension/Sub-dimension	Maturity level	Maturity score
Smart products and services	Level 2: Survival	2.50
Smart business processes	Level 1: Existence	1.60
Smart production and operations	Level 1: Existence	0.95
Smart marketing and sales operations	Level 1: Existence	1.55
Supportive operations	Level 2: Existence	2.29
Strategy and organization	Level 1: Existence	1.14

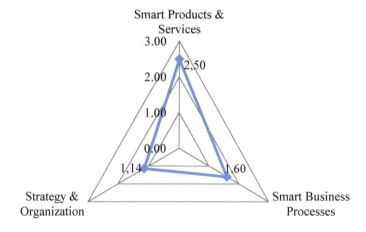

Fig. 4.1 Maturity levels of dimensions of company in a radar chart

4.6   Conclusion

The study presented here aimed to develop of an Industry 4.0 maturity model and an assessment survey to provide companies a tool to help them understand their current state regarding to Industry 4.0. Different application areas were proposed for Industry 4.0 such as smart finance, smart marketing and human resources in order to differentiate the model and increase companies' perspective for Industry 4.0 applications. Future studies will aim at diversifying Industry 4.0 maturity model to enhance the industry scope with weighted associated fields for each industry and create activity plans according to companies' current maturity level.

Appendix: Survey Questionnaire

Smart Products and Services

Principles	Technologies
Real time data management (Collection/Processing/Analysis/Inference) Interoperability Decentralized Service oriented	Data analytics and Artificial intelligence Embedded systems Communication and Networking Cybersecurity Sensors and Actuators Cloud RFID and RTLS Technologies

Questionnaire
1. Which functions can your company's products fulfill the following options?

Communicating with other products/platforms, machines and external systems
Collecting data from environment and other systems
Keeping the data they collect on their system or in the cloud
Having a platform on which the product or cloud applications are working

2. What stages of the data analysis can the product perform? (Porter and Heppelmann 2015)

Descriptive—Capture products' condition, environment and operation
Diagnostic—Examine the causes of reduced product performance or failure
Predictive—Detect patterns that signal impending events
Prescriptive—Identify measures to improve outcomes or correct problems

3. To what extent can products be tracked throughout their lifecycle? (The University of Warwick Maturity Model)

No or limited product tracking
Products can be tracked as they move between manufacturing and internal distribution sites
Products can be tracked through manufacturing and distribution until they reach the customers DC
Products can be tracked along their complete lifecycle

4. Who do you offer service/insights for according to the user data obtained from the product?

None
Business
Customers
Partners

Smart Business Processes
Production, Logistics and Procurement

Principles	Technologies
Real time data management (Collection/Processing/Analysis/Inference) Interoperability Virtualization Decentralized Agility Integrated business processes	Data analytics and Artificial Intelligence Adaptive robotics Simulation Communication and Networking Cybersecurity Additive manufacturing Virtualization technologies Sensors and Actuators Cloud RFID and RTLS Technologies Mobile Technologies

Questionnaire

1. Which of the following systems do you use? Does the system have an interface to the leading system? (Lichtblau et al. 2015)

	Interface to leading system	
	No	Yes
MES—manufacturing execution system		
ERP—enterprise resource planning		
PDM—product data management		
PPS—production planning system		
PDA—production data acquisition		
MDC—machine data collection		
CAD—computer-aided design		
SCM—supply chain management		

2. To what extent is the current supply chain integrated? (The University of Warwick Maturity Model)

Ad hoc reactive communication with suppliers and customers
Basic communication and data sharing where required with suppliers and customers
Data transfer between key strategic suppliers/customers (for example customer inventory levels)
Fully integrated systems with suppliers/customers for appropriate processes (for example real time integrated planning)

3. To what extent are the production equipment and systems automated?

Machine level: Partial
Machine level: Exact (Loading/Unloading + Operation)
Production line/cell level: Partial
Production line/cell level: Exact (Loading/Unloading + Operation +Transportation)
Factory level: Partial

4. Express the level of personalization in production.

Low—10,000 + batch size
Medium
High—1 batch size

5. Which data about your machinery, processes, and products as well as malfunctions and their causes is collected during production, and how is it collected? (Lichtblau et al. 2015)

	Manually	Automatically
Inventory data		
Manufacturing throughput times		
Equipment capacity utilization		
Production residues		
Error quota		
Employee utilization		
Data on remaining processing		
Overall equipment effectiveness (OEE)		
Other:		

6. How is the data you collect used in production? (Lichtblau et al. 2015)

Predictive maintenance
Optimization of logistics and production processes
Creation of transparency across production process
Quality management
Automatic production control through use of real-time data
Optimization of resource consumption (material, energy)
Other:

7. How is the data you collect used in logistics and procurement? (Schreiber et al. 2016)

Predictive supplier risk management (to detect supplier failures early on)
Digital supplier scorecards, objectives and improvement tracking.
Automated tracking of target achievement and bonus payments
Digital claim management system with integrated automatic warning system
Big data analytics to detect new suppliers globally

8. To what extent does your supply chain an end-to-end visibility? (The University of Warwick Maturity Model)

No integration with suppliers or customers
Site location, capacity, inventory and operations are visible between first tier suppliers and customers
Site location, capacity, inventory and operations are visible throughout supply chain
Site location, capacity, inventory and operations are visible in real time throughout supply chain and used for monitoring and optimization

9. What is the level of real-time traceability of the operation in the digital environment? (Digital-twin concept)

None
Machine level
Production line/cell level
Factory level

10. What is the use level of technologies in production, logistics and procurement?

	Mobile and virtual technologies	3D Printers	Adaptive and collaborative robots
None			
Low			
Medium			
High			

Smart Business Processes
R&D—Product Development

Principles	Technologies
Real time data management (Collection/Processing/Analysis/Inference) Virtualization Agility	Data analytics and Artificial intelligence simulation communication and Networking Cybersecurity additive manufacturing virtualization technologies cloud RFID and RTLS technologies

Questionnaire

1. To what extent are the manufacturability and terms of use of the product simulated during product development?

None
Low
Medium
High

2. To what extent is the data obtained from the product used in the new product development?

None
Low
Medium
High

3. Do you use 3D printers in the production/prototyping processes?

No
Yes

4. Is product design information automatically transferred with the CAD/CAM systems to the machine?

No
Yes

5. Can your customers customize your products before production according to their preferences?

No
Yes

Smart Business Processes
After Sales Services

Principles	Technologies
Real time data management (Collection/Processing/Analysis/Inference) Virtualization Agility Service oriented	Data analytics and Artificial intelligence Embedded systems Communication and Networking Cybersecurity Virtualization technologies Cloud RFID and RTLS technologies Mobile technologies

Questionnaire
1. How do you benefit from data you collect in after-sales services?

Early detection of product quality issues and focused recalls
Improved product design
Advanced supplier recovery
Optimized spare parts planning
Minimized suspect and fraudulent claims
Reduced "remorse returns" and no trouble found rates
Increased reserves forecast accuracy
Enhanced service quality and service information
Intensified customer intimacy and next best action

2. Which services do you provide by using data analytics and other technologies in after-sales services?

Remote maintenance
Assistance with problems or faults in real time
IT-assisted claim management
Order management (CRM, order history, delivery tracking, etc.)
Display of product history
Delivery forecast

3. Do you utilize from digital technologies (mobile and virtualization technologies) in after-sales service processes?

No
Yes

Smart Business Processes
Pricing/Promotion

Principles	Technologies
Real time data management (Collection/Processing/Analysis/Inference)	Data analytics and Artificial intelligence
Decentralized	Communication and Networking
Service oriented	Cybersecurity
Integrated business processes	Cloud

Questionnaire
1. Which of the following studies are conducted within customer analytics?

Customer segmentation
Customer lifetime value
Cross selling
Campaign management
Market basket analysis/product bundling
Product recommendation
Customer churn analysis
Product portfolio management

2. Do you utilize from data obtained from environment/other platforms in product pricing or dynamic pricing?

	Product pricing	Dynamic pricing
No		
Yes		

3. Do you generate new campaigns from purchasing and product usage data?

No
Yes

4. Do campaign management systems work integrated with other systems?

No
Yes

5. Do you analyze campaign performance to use these analyses in new campaigns?

No
Yes

Smart Business Processes Sales and Distribution Channels

Principles	Technologies
Real time data management (Collection/Processing/Analysis/Inference) Agility Service oriented	Data analytics and Artificial Intelligence Communication and Networking Cloud Mobile technologies

Questionnaire
1. What is the level of sales team support with digital products and services and real-time access to systems?

None
Low
Medium
High

2. Do you conduct real-time profitability analysis?

No
Yes

3. Do you use real-time and automated performance management systems for local sales force?

No
Yes

4. To what extent are your sales channels integrated?

None
Low
Medium
High

5. To what extent do you use integrated channels to communicate with customers and to manage customer interaction?

None
Low
Medium
High

6. To what extent do you collaborate with partners to reach customers (i.e. exchange of customer insight, etc.)?

None
Low
Medium
High

7. Which content analyses are performed on social media?

None
Sentiment analysis
Trend analysis

Smart Business Processes
Human Resources

Principles	Technologies
Real time data management (Collection/Processing/Analysis/Inference) Agility	Data analytics and Artificial intelligence Cloud Mobile technologies

Questionnaire

1. In what areas is the data collected and data analytics is used?

	Data collected	Data analytics used
Capability analytics—(a talent management process that allows you to identify the capabilities or core competencies you want and need in your business.)		
Capacity analytics—(seeks to establish how operationally efficient people are in a business.)		
Competency acquisition analytics—(the process of assessing how well or otherwise your business acquires the desired competencies.)		
Employee churn analytics—(the process of assessing your staff turnover rates in an attempt to predict the future and reduce employee churn.)		
Corporate culture analysis—the process of assessing and understanding more about your corporate culture or the different cultures that exists across your organization.)		
Recruitment channel analytics—(the process of working out where your best employees come from and what recruitment channels are most effective.)		
Leadership analytics—(unpacks the various dimensions of leadership performance via data gained through the use of surveys, focus groups, employee interviews or ethnography.)		
Employee performance analytics—(seeks to assess individual employee performance.)		

2. Can your company share real-time data with employees in the field?

No
Yes

3. Can employee training be carried out in a virtual environment?

No
Yes

Smart Business Processes Information Technology

Principles	Technologies
Real time data management (Collection/Processing/Analysis/Inference) Interoperability Virtualization	Data analytics and Artificial intelligence Communication and Networking Cybersecurtiy

(continued)

(continued)

Principles	Technologies
Decentralized	Cloud
Integrated business processes	Mobile technologies

Questionnaire

1. How far along are you with your IT security solutions? (Lichtblau et al. 2015)

	Solution planned	Solution in progress	Solution implemented
Security in internal data storage			
Security of data through cloud services			
Security of communications for in-house data exchange			
Security of communications for data exchange with business partners			

2. Are you already using cloud services? (Lichtblau et al. 2015)

	Cloud-based software	For data analysis	For data storage
Production, Logistics and Procurement			
R&D—Product development			
After sales services			
Sales and Distribution channels			
Pricing/Promotion			
Human resources			
Information technology			
Finance			

3. Do IT dashboards be used for traceability of company processes?

No
Yes

4. How would you evaluate your equipment infrastructure when it comes to the following functionalities? (Lichtblau et al. 2015)

	No, not available	Yes, to some extent	Yes, completely
Machines/systems can be controlled through IT			
M2 M: machine-to-machine communications			
Interoperability: integration and collaboration with other machines/systems possible			

Smart Business Processes
Smart Finance

Principles	Technologies
Real time data management (Collection/Processing/Analysis/Inference) Decentralized	Data analytics and Artificial intelligence Cloud

Questionnaire

1. Do you perform real-time cost calculations with data obtained from production?

No
Yes

2. Do you analyze company's cash flow and investments on a historical basis?

No
Yes

3. To what extent do you utilize from financial data when make investment decision?

None
Low
Medium
High

4. To what extent are your financial systems automated?

None
Low
Medium
High

5. How do you perform financial risk measurement?

None
Historical basis
Real-time

Strategy and Organization
Business Models
Questionnaire
1. Do your existing products and services comply with innovative digital business models?

No
Yes

2. To what extent are you aware of the "As-a-service" business model? (The University of Warwick Maturity Model)

No awareness.
Aware of concept with some initial plans for development
High awareness and implementation plans are in development
"As-a-service" has been implemented and is being offered to the customer

3. Which degree of resource is allocated to digital business models?

None
Low
Medium
High

4. Is the current business model of the company evaluated and updated during the interim period in the matter of digitization?

No
Yes

5. To what extent do you monetize your new data-driven services?

None
0–2.5%
2.5–10%
Over 10%

Strategy and Organization
Strategic Partnerships
Questionnaire

1. Does your company have partnerships for Industry 4.0 projects with following options?

None
Academics
Technology providers
Suppliers
Customers

2. How would you describe the implementation status of your Industry 4.0 strategy? (Lichtblau et al. 2015)

No strategy exists
Pilot initiatives launched
Strategy in development
Strategy formulated
Strategy in implementation
Strategy implemented

3. Do you use indicators to track the implementation status of your Industry 4.0 strategy?

No, our approach is not yet that clearly defined
Yes, we have a system of indicators that gives us some orientation
Yes, we have a system of indicators that we consider appropriate

Strategy and Organization
Technology Investments
Questionnaire

1. Which technologies in your company are driving Industry 4.0?

None
Data analytics and Artificial intelligence
Adaptive robotics
Simulation
Embedded systems
Communication and Networking
Cybersecurity
Cloud
Additive manufacturing

(continued)

(continued)

Virtualization technologies (VR & AR)
Sensors and Actuators
RFID and RTLS technologies
Mobile technologies

2. To what extent do you allocate sufficient budget to investments in Industry 4.0?

None
Low
Medium
High

3. How often do you conduct a cost/benefit analysis for Industry 4.0 investment? (The University of Warwick Maturity Model)

No measurable Industry 4.0 investment yet
No ongoing review of cost/benefit analysis for Industry 4.0 investment yet
Annual cost/benefit analysis of Industry 4.0 investment
Quarterly cost/benefit analysis of Industry 4.0 investment

4. In which parts of your company have you invested in the implementation of Industry 4.0? (Lichtblau et al. 2015)

	Planning investment	Investment done
Production, Logistics and Procurement		
R&D—Product development		
After sales services		
Pricing/Promotion		
Sales and Distribution channels		
Human resources		
Information technology		
Finance		

Strategy and Organization
Organizational Structure and Leadership
Questionnaire
1. Are business units/project teams structured in interdisciplinary in the company?

No
Yes

2. Is there any business unit to maintain relationship or communicate with customers?

No
Customer service
Customer relationship management

3. Is there any data-driven organizational structure? (Data scientists, analytics team, digital transformation director, etc.)

No
Yes

4. To what extent are employees equipped with relevant skills for Industry 4.0? (The University of Warwick Maturity Model)

Employees have little or no experience with digital technologies
Technology focused areas of the business have employees with some digital skills
Most areas of the business have well developed digital and data analysis capability
All across the business, cutting edge digital and analytical skills are prevalent

5. Do you have training for the digital transformation in the company?

No
Yes

6. How is your IT organized? (Lichtblau et al. 2015)

No in-house IT department (service provider used)
Central IT department
Local IT departments in each area (production, product development, etc.)
IT experts attached to each department

7. To what extent do departments collaborate with each other? (The University of Warwick Maturity Model)

The business operates in functional silos
There is limited interaction between departments (i.e. S&OP process)
Departments are open to cross-functional collaboration
Departments are open to cross-company collaboration to drive improvements

8. To what extent does the leadership team support Industry 4.0? (The University of Warwick Maturity Model)

Leadership team does not recognize the value of the Industry 4.0 investments
Leadership team is investigating potential Industry 4.0 benefits
Leadership team recognizes the financial benefits to be obtained through Industry 4.0 and is developing plans to invest
Widespread support for the Industry 4.0 within both the leadership team and across the wider business

9. How is your Industry 4.0 team organized to execute innovative projects?

There is no employee for Industry 4.0 projects
There are employees for Industry 4.0 project; but in different business units
There are employees for Industry 4.0 project in the same business unit

10. Is there any working environment where OT/IT units work together?

No
Yes

References

Backlund F, Chronéer D, Sundqvist E (2014) Project management maturity models–A critical review: a case study within Swedish engineering and construction organizations. Procedia-Social and Behavioral Sciences 119:837–846

Duffy J (2001) Maturity models: blueprints for e-volution. Strategy and Leadership 29(6):19–26

Erol S, Schumacher A, Sihn W (2016) Strategic guidance towards Industry 4.0–a three-stage process model. In International conference on competitive manufacturing

Ganzarain J, Errasti N (2016) Three stage maturity model in SME's toward industry 4.0. J Ind Eng Manag 9(5):1119

Geissbauer R, Vedso J, Schrauf S (2016) Industry 4.0: Building the digital enterprise. Retrieved from PwC Website: https://www.pwc.com/gx/en/industries/industries-4.0/landing-page/industry-4.0-building-your-digital-enterprise-april-2016.pdf

Lichtblau K, Stich V, Bertenrath R, Blum M, Bleider M, Millack A,... Schröter M (2015) IMPULS-Industrie 4.0-Readiness. Impuls-Stiftung des VDMA, Aachen-Köln

Mettler T (2009) A Design Science Research Perspective on Maturity Models in Information Systems. Working Paper. Institute of Information Management, Universtiy of St. Gallen, St. Gallen

Nikkhou S, Taghizadeh K, Hajiyakhchali S (2016) Designing a portfolio management maturity model (Elena). Procedia-Social and Behavioral Sci 226:318–325

Porter ME, Heppelmann JE (2015) How smart, connected products are transforming companies. Harvard Bus Rev 93(10):96–114

Proença D, Borbinha J (2016) Maturity models for information systems-A state of the art. Procedia Comput Sci 100:1042–1049

Retrieved from https://i40-self-assessment.pwc.de/i40/interview/

Retrieved from https://warwickwmg.eu.qualtrics.com/jfe/form/SV_7O3ovIWlTCu90uF

Rockwell Automation. (2016). The Connected Enterprise Maturity Model. Retrieved from Website:http://literature.rockwellautomation.com/idc/groups/literature/documents/wp/cie-wp002_-en-p.pdf

Schmidt R, Möhring M, Härting RC, Reichstein C, Neumaier P, Jozinović P (2015) Industry 4.0-potentials for creating smart products: empirical research results. In International conference on business information systems springer international publishing.:16–27

Schreiber B, Janssen R, Weaver S, Peintner S (2016) Procurement 4.0 in the digital world. Retrieved from Website: http://www.adlittle.com/downloads/tx_adlreports/ADL__Future__of__Procurement__4.0.pdf

Schumacher A, Erol S, Sihn W (2016) A maturity model for assessing industry 4.0 readiness and maturity of manufacturing enterprises. Procedia CIRP 52:161–166

Tarhan A, Turetken O, Reijers HA (2016) Business process maturity models: A systematic literature review. Inf Softw Technol 75:122–134

Chapter 5
Technology Roadmap for Industry 4.0

Peiman Alipour Sarvari, Alp Ustundag, Emre Cevikcan, Ihsan Kaya
and Selcuk Cebi

Abstract From both strategic and technologic perspectives, the Industry 4.0
roadmap visualizes every further step on the route towards an entirely digital
enterprise. In order to achieve success in the digital transformation process, it is
necessary to prepare the technology roadmap in the most accurate way. The intent
of this chapter is to present a technology roadmap for Industry 4.0 transformation to
facilitate the planning and implementation process.

5.1 Introduction

Technology road mapping is an important method that has become integral to creating and delivering strategy and innovation in many organizations. The graphical
and collaborative nature of roadmaps supports strategic alignment and dialogue
between functions in the firm and between organizations (IFM 2016). The technology
roadmap process addresses the identification, selection, acquisition, development,
exploitation, and protection of technologies (product, process, and infra structural)
needed to achieve, maintain and grow a market position and business performance
matching with the company's objectives (Toro-jarrín et al. 2016).

Rob Phaal is the pioneer and first developer of 'T-plan technology road mapping' in the 2000s (Phaal et al. 2001). This original work dedicated to the new
methodology of taking a market-pull strategy, and gives a step-by-step framework
on how to utilize road mapping in firms by using minimal resources. Consequently,
his work became a primary framework for road mapping for both market pull and
technology push approaches (Phaal et al. 2001). His approach is a tool for managers

P.A. Sarvari (✉) · A. Ustundag · E. Cevikcan
Department of Industrial Engineering, Faculty of Management,
Istanbul Technical University, Macka, Istanbul, Turkey
e-mail: alipoursarvarip@itu.edu.tr

I. Kaya · S. Cebi
Department of Industrial Engineering,
Yildiz Technical University, Besiktas, Istanbul, Turkey

© Springer International Publishing Switzerland 2018
A. Ustundag and E. Cevikcan, *Industry 4.0: Managing The Digital Transformation*,
Springer Series in Advanced Manufacturing, https://doi.org/10.1007/978-3-319-57870-5_5

and firms' policy makers to develop a roadmap very promptly, gives a chance to combine the development of technologies and activities for their exploitation and commercialization. However, many enterprises are incapable of launching roadmaps due to a lack of qualified staff for this process. In 2004, Phaal concluded that a qualified specialist in long-term planning horizon should handle the road mapping process (Phaal et al. 2004a).

Daim and Oliver (2008) define roadmap as a way to identify and decide upon trajectories to follow to reach future success, likewise as a traditional map guides travelers to their destination. The underlying principals usually relate to three specific characteristics, as it provides the corporation with an illustration of the current state, a desirable future state, and strategies to reach the future state (Phaal et al. 2004a). The key benefits of technology road mapping are given as follows:

- Establishing alignment of commercial and technical strategies
- Improving communication across teams and organizations
- Examining potential competitive strategies and ways to implement those strategies
- Efficient time management and planning
- Identifying the gaps between technology, market, and product intelligence
- Prioritizing the investments
- Setting competitive and rational targets
- Guiding and leading the project teams
- Visualizing outputs including goals, processes, and progresses.

Phaal et al. (2004b) identify eight types of graphical roadmaps according to the specific needs each one is a useful and recursive tool for strategic management. For study purpose, it was chosen multiple layers type, which is the most common format for a technology roadmap for the study of the current and future states related to three primary levels marked by Cosner et al. (2007): market, product, and resources. Market level describes current, and future customer needs accompanying with competitive strategies, regulatory environment, complementary product evolution, substitute products, disruptive innovations, and other factors. Strategic goals are stated as milestones or target dates for specific events; Product level documents performance and product features' evolution, new-to-the-company products (including services) and new-to-the-world products, and; Technology level describes expected R&D products, their availability dates, the driving factors for the R&D, and related information (Toro-jarrín et al. 2016).

Technology roadmaps can be used to facilitate the co-ordination of different staff functions. For instance, the widespread distrust between R&D and marketing units may be mitigated by means of a conjoint roadmap to which R&D contributes technological factors while marketing staff brings in a product-related perspective. Another field in which roadmaps prove to be helpful is that of competitive strategy. In this case, the interest groups involved are the company's marketing unit and its customers. The strategic utilization of roadmaps is principally reflected by the organization's announcement policy. A striking example of this can be recognized in the computer industry, where the Microsoft Corporation regularly succeeds in deterring consumers from purchasing a competitive product by explicitly announcing the upcoming launch of a similar article (so-called 'vaporware'). Furthermore, technology roadmaps enable the coordination of inter- and extra corporate R&D activities. This function especially presents itself where huge cooperation or high levels of external procurement are expected. Eventually, individual companies have the option to join forces in devising a technology roadmap that supports their common orientation. This has happened very prominently in the semiconductor industry (Erol et al. 2016).

In today's business, Industry 4.0 is driven by digital transformation in vertical/horizontal value chains and product/service offerings of the companies. The required key technologies for Industry 4.0 transformation such as artificial intelligence, internet of things, machine learning, cloud systems, cybersecurity, adaptive robotics cause radical changes in the business processes of organizations. The challenges for Industry 4.0 transformation are determined as:

- Lack of knowledge about technologies and their opportunities
- Uncertainty about the benefits of technology investments on products and processes
- Lack of knowledge about customer demand regarding new products and business models under industry 4.0 vision
- Limited human and financial resources
- Difficulties to spot the starting point and milestones of planning horizon
- Need for efficient portfolio management for technology investments
- Requirements for prioritization and scheduling of new product and process projects
- Allocating the limited resources to the projects and collaborating with reliable partners
- Lack of communication about the benefits of the Industry 4.0 transformation projects through the organization.

Therefore, the strategy followed by a company while adopting new technologies to its processes and products has a critical importance. As the first step, the organization has to develop a roadmap which is a complex long-term planning instrument that allows for setting strategic goals and estimating the potential of new technologies, products, and services (Vishnevskiy et al. 2016).

The objective of this study is to facilitate to managers, policy makers and practitioners the creation of a technology roadmap for Industry 4.0 transformation. This chapter proposes a comprehensive framework for Industry 4.0 road mapping. The main goal is to overcome the challenges and difficulties confronted by the companies in the digital transformation process.

5.2   Proposed Framework for Technology Roadmap

In a promptly changing world, to visualize the future is not only an additional tool for strategic planning but a necessary practice for every company. Rohrbeck and Schwarz (2013) emphasized that the early acknowledge and visionary anticipation of the technological potential play a pivotal role in a business environment which is characterized by the improvement of competitiveness. They also observed that ignoring changes in a globalized world often results in losing opportunities or failing in responding threats. For Makridakis (Godet 2010) the role of visioning the future is to provide managers and government policy maker's different ways to comprehend the future and help them have a whole perception of possible implications of social and technological changes.

A "roadmap" empowers whoever in the industry to simply understand each move and what decisions need to be done, who needs to make them and when. This procedure is decoded into a project plan, specifying the characteristics of work in each of the associated stages of formation. In our proposed framework in Fig. 5.1, the strategies and key technologies are defined in the first phase of the roadmap, subsequently, in the second phase, development of the new products and processes is conducted.

In the new product and process development phase, ideas are generated at first, the generated ideas are evaluated and some potential ideas are selected for the implementation. Considering the natural constraints like budget limits, partnership policies and inadequate human resources just some of the chosen ideas can be projected in shape of one or several portfolios. Project prioritization usually has been conducted based on its benefits and added values. Scheduling which is the time-horizoned framework of Industry 4.0 roadmap has visualized the technologies, targets, processes and progress levels. The last step of road mapping is the implementation, which is settling down the defined product and process projects, and dynamically reviewed compared to the updated ideas, new products, new key technologies and better processes. All steps of proposed road mapping flow are going to be described in details via following sections.

5.2.1 Strategy Phase

A strategy is defined in a time-based plan and portrays where an industry is, where it needs to go, and how to get it there. Strategy phase of the roadmap is a collaborative procedure to planning. In the strategy phase, the evaluation of the enterprise digital maturity to set clear targets for the next years (based on the time horizon) is the first step that one should take into account during preparing the Industry 4.0 roadmap. Many industrial capabilities have already begun digitizing their business, but often the process has started in the low levels of the organization. Taking the time to evaluate the maturity level in all areas of Industry 4.0 can help to understand what strengths are possible to build already on, and which systems/processes are needed to combine into future solutions. Simultaneously to think about where an organization wants to go in the future, one should consider what the organization could gain by co-operating with clients, suppliers, technology partners and even competitors, without restricting the vision based on current limitations. The focus should be beyond the technical details and estimates the

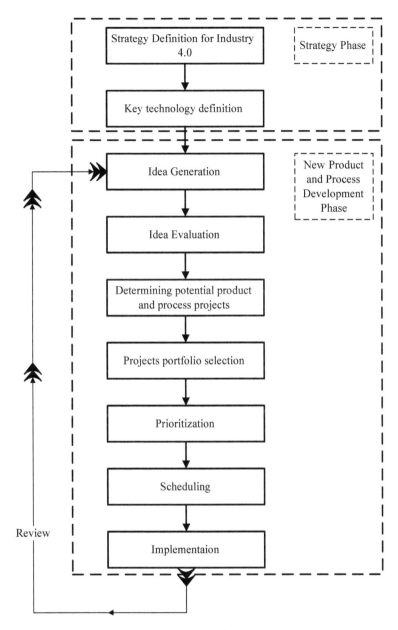

Fig. 5.1 Proposed technology roadmap for Industry 4.0

impacts of new applications on organization's value chain. The roadmap will need to consider future changes in customer behavior and how the organization relationship with them will change. Driving from the current to the future desired state will demand precise steps and a clear prioritization.

Strategies answer the question of what has to be done to achieve the desired outcome (Osterwalder and Pigneur 2010). From another side, the technology development and market demands accelerate changes in the world. From the strategic perspective, it is frequently found that technology innovation is a key advantageous factor for performance improvement and survival of the enterprise, besides it is a determinant factor for the sustainable economic growth of nations and quality life improvement of their people (Keupp et al. 2012). According to Rohrbeck and Schwarz (2013), firms suffer blindness caused by focusing mostly on the inside of the company and reinforcing practices that made the company successful in the past thus, it becomes evident that firms need to dedicate efforts to look outside the company and to be aware of the coming changes. In another word, the capacity for bringing value to the customer is strongly related to the technology development. Furthermore, before new product and process development phase, the key technologies having strategic importance for the company should be determined meticulously.

5.2.2   New Product and Process Development Phase

The new product and process development phase allows sketching goals and projects considering different principles and constraints on separate layers against a shared vision. For our purpose, we use a distinction into three technical perspectives: technology constraints (budget and partnerships), goals (saving, revenue, risks) and projects. Three layers represent these perspectives. Figure 5.2 depicts the three layers of portfolio selection and project prioritization.

At this moment, the generated ideas are selected by experts based on the products and processes feasibility dimensions. After listing the potential projects, the selection stage of the projects portfolios is fulfilled. Prioritization stage is conducted considering saving and risk factors for new process development projects, also the revenue and risk factors for new product development projects. Figure 5.2 illustrates a flow of matrices in terms of project selection and forming portfolios.

It is worth to note that the optimal portfolio should lie on the Efficient Frontier curve which constitute saving and revenue as the vertical dimension whereas the horizontal dimension reflects the risk. The chart shown on Fig. 5.3 demonstrates how the optimal portfolio forms. The optimal risk portfolio is normally defined to be around in the center of the curve since as one goes higher up the curve, one takes on proportionately more risk for a lower incremental return. Oppositely low risk/low return portfolios are pointless since you can gain a similar return by investing in risk-free assets. You can pick how much volatility you can bear in your portfolio by selecting any other point that falls on the efficient frontier. So your true selected portfolio will give you the maximum saving and revenue for the percentage of risk you are planning to accept. Briefly, the portfolios above the efficient frontier curve are impossible and the portfolios below the efficient frontier are not efficient because for the same risk one could achieve a greater return.

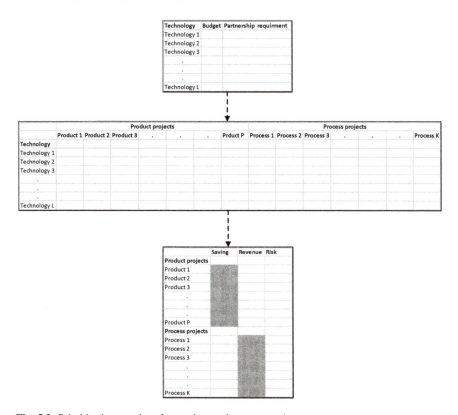

Fig. 5.2 Prioritization matrices for product and process projects

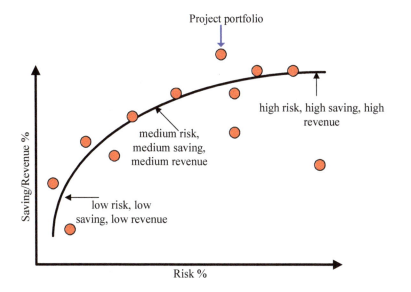

Fig. 5.3 Efficient frontier curve for portfolio selection

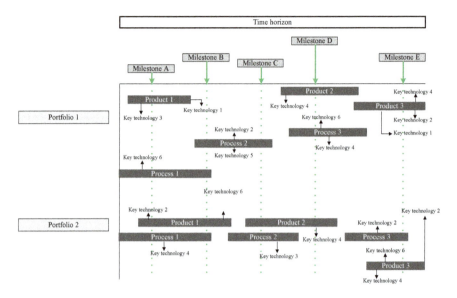

Fig. 5.4 An exemplary scheduled Industry 4.0 roadmap for two portfolios

Scheduling is the next stage of the mapping defining milestones and projects statuses. The output of this stage is a timely ordered and multi-perspective map of the overall approach towards the envisioned Industry 4.0 concept that builds the strategic frame for concrete actions. The time horizon for the accomplishment of the Industry 4.0 vision settles whether a company takes a revolutionary strategy versus an evolutionary strategy. Figure 5.4 illustrates an exemplary scheduled roadmap for twelve projects in the shape of two portfolios with specified milestones and engaged key technologies in a strategically defined time horizon.

Each box in the roadmap represents a state or project on the way to the ultimate Industry 4.0 vision. Additionally as well, a strategy to reach the goal is shaped into milestones. Milestones are progressed percentages of savings, budget usages, terminated projects, applied technologies, etc. The required key technologies represented on each box of projects. The time dimension is indicated by a rough time-frame or a concrete date depending on the overall time horizon of the company.

5.3 Conclusion

Considering a globalizing world, the need to implement development strategies that can sustainably guarantee the competitiveness of companies is the main issue. It is in this context that Industry 4.0 roadmap presents itself as a visually pictured clear path to boost the competitiveness of companies. The Industry 4.0 roadmap gives an overview of the current situation of the company and the perspective situation to be reached in a time horizon. A genuine prediction of Industry 4.0 has created unique ideas shaping products, processes, and opportunities in terms of defined targets and milestones. In this regard, the continuously evolving key technologies and solutions are an undeniable part of fulfilling the vision of the designed roadmaps. This chapter tried to present a comprehensive Industry 4.0 roadmap framework covering the strategy, new product and process development phases. The proposed roadmap design flow covers a broad range of strategic planning to implementation of the defined visions to facilitate building manufacturers' Industry 4.0 roadmaps. With reminding the criticality of review and feedback concepts of the dynamic structure of the cycle between idea generation stage and the implementation step, the chapter concludes that it is important that company managers understand and follow presented above core principles for not facing with difficult replacement decisions.

References

Cosner RR, Hynds EJ, Fusfeld AR, Loweth CV, Scouten C, Albright R (2007) Integrating roadmapping into technical planning. Res Technol Manag 50(6): 31–48

Daim T, Oliver T (2008) Implementing technology roadmap process in the energy services sector: a case study of a government agency. Technol Forecast Soc Chang. 75:687–720

Erol S, Schumacher A, Sihn W (2016) Strategic guidance towards Industry 4.0—a three-stage process model. In: International conference on competitive manufacturing

Godet M (2010) Future memories. Technol Forecast Soc Chang 77:1457–1463

Keupp MM, Palmié M, Gassmann O (2012) The strategic management of innovation: a systematic review and paths for future research. Int J Manag Rev 14:367–390

Osterwalder A, Pigneur Y (2010) Business model generation: a handbook for visionaries, game changers, and challengers. Wiley, New York

Phaal R, Farrukh C, Probert D (2001) T-plan: the fast start to technology roadmapping: planning your route to success. Institute for Manufacturing, University of Cambridge, Cambridge, UK

Phaal R, Farrukh C, Probert D (2004a) Customizing roadmapping. Res Technol Manag 47(2):26–37

Phaal R, Farrukh CJ, Probert DR (2004b) Technology roadmapping—a planning frame-work for evolution and revolution. Technol Forecast Soc Chang 71:5–26

Roadmapping at IfM—Institute for Manufacturing, University of Cambridge. Retrieved from http://www.ifm.eng.cam.ac.uk/ifmecs/business-tools/roadmapping/roadmapping-at-if

Rohrbeck R, Schwarz JO (2013) The value contribution of strategic foresight: insights from an empirical study of large European companies. Technol Forecast Soc Chang 80:1593–1606

Toro-jarrín MA, Ponce-jaramillo IE, Güemes-castorena D (2016) Technological Forecasting & Social Change Methodology for the of building process integration of Business Model Canvas and Technological Roadmap. Technol Forecast Soc Chang 110:213–225

Vishnevskiy K, Karasev O, Meissner D (2016) Technological Forecasting & Social Change Integrated roadmaps for strategic management and planning. Technol Forecast Soc Chang 110:153–166

Chapter 6
Project Portfolio Selection for the Digital Transformation Era

Erkan Isikli, Seda Yanik, Emre Cevikcan and Alp Ustundag

Abstract By bringing various technological advances together, Industry 4.0 promises production systems to boost in productivity. The transformation of the existing production systems into an Industry 4.0 factory is a strategic and long-term undertaking which needs a high capital investment, training of personnel and change of the environment and the culture in almost all of the functions of the value chain. Thus, it is required to be planned well by the companies. To be successful in the Industry 4.0 era, companies will not only have to simultaneously execute interdependent projects, but also require to perform project portfolio selection task in multi-dimensional environments in the presence of high uncertainty. Since determining optimal project portfolio for digital transformation among various project alternatives demands the consideration of multiple constraints and inter-dependencies, an integer programming model is proposed to address this problem in this chapter. To demonstrate its effectiveness and practicality, the proposed optimization model is applied to the Industry 4.0 project alternatives of an automotive manufacturer.

Abbreviations

PPM   Project Portfolio Management
PPOM Project Portfolio Optimization and Management
LP   Linear Programming

E. Isikli (✉) · S. Yanik · E. Cevikcan · A. Ustundag
Department of Industrial Engineering, Faculty of Management, Istanbul Technical University, 34367 Macka, Istanbul, Turkey
e-mail: isiklie@itu.edu.tr

S. Yanik
e-mail: sedayanik@itu.edu.tr

E. Cevikcan
e-mail: cevikcan@itu.edu.tr

A. Ustundag
e-mail: ustundaga@itu.edu.tr

© Springer International Publishing Switzerland 2018
A. Ustundag and E. Cevikcan, *Industry 4.0: Managing The Digital Transformation*,
Springer Series in Advanced Manufacturing, https://doi.org/10.1007/978-3-319-57870-5_6

6.1 Introduction

It is important to manage technology investments for the survival of a company to cope with high competition, demanding customers, reduced lead times and need for higher flexibility in today's business environment. Industry 4.0 promises autonomous production systems in smart factories resulting in increased efficiencies. Man, machine and resources can directly communicate with each other, products are smart and know how they will be manufactured and the whole value chain is integrated. To achieve this envisioned future state of the manufacturing environment, traditional industry needs to undertake a structured transformation program whereas Industry 4.0 transition has many technical, economical, organizational and legal challenges. An effective management of project portfolio to undertake these challenges in the Industry 4.0 transition can be a key initiative for the future success of a company.

Project portfolio management (PPM) goes beyond project management by not only focusing on doing projects right but also doing the right projects. Organizations generally undertake multiple projects. Selecting and prioritizing the set of projects are crucial because most of the projects are related through interdependencies and they need to share scarce resources. Besides, a strategic perspective is needed because projects serve as a means to achieve strategy implementation of organizations. A strategic PPM is expected to go beyond prioritization, selection and evaluation of projects and achieve the required structural alignment and organizational transformations. The issues considered during PPM are generally the evaluation of strategic fit of all projects within a company, analysis of unpredictability to the company represented as risks, assessment of market (financial) potential, changes in company operations, modelling interdependencies between a set of projects, identifying resource constraints shared between projects, selection and prioritization of projects and optimization of the portfolio (Pinto 2016; De Reyck et al. 2005; Kaiser et al. 2015).

PPM aims to maximize the economic benefit of the new technology investments within the framework of Industry 4.0 while considering technical, organizational and legal challenges. It can be compared to fund investment management in the stock market where the goal is to maximize the value of the portfolio as well as return on the spending. A well-balanced portfolio is mainly characterized by finances together with a consideration of specialized staff, equipment and buildings, intellectual property, a budget, and a host of inherited relationships with different units and functions of the company. Industry 4.0 symbolizes the revolution the traditional industry needs to make using the advanced technology and innovative business models especially in the manufacturing systems. Some other terms used for Industry 4.0 are internet of things, cyber physical systems, smart systems and digital factory. This can only be enabled by the application of various technologies which will ensure real-time information access, data integration, process flexibility, predictive maintenance and security. Such a comprehensive transformation of the on-going manufacturing environment can be achieved by an effective program management of projects because these projects share common resources, have interdependencies and precedence relationships as well as different risk and benefit potentials. Thus, an optimized project portfolio is required to model these complex relationships between the projects, the project and the organization/environment to achieve the best financial output under uncertainties.

In this study, we use the three main stages of the PPM approach owed from Jeffery and Leliveld (2004): (i) definition of the projects, (ii) management of the projects, (iii) optimization of the portfolio. In the definition step, we define the projects that could be required by a traditional company for the Industry 4.0 transition. Then, a high-level cost and benefit estimation and precedence relationships between the projects are done. In the project management step, detailed risk, cost and financial investment measures are quantified as well as constraints and limitations are specified. Subsequently, in the optimization step, we balance and optimize the Industry 4.0 project portfolio. We propose a simulation-based optimization approach in which uncertainties can be represented using probability distribution and incorporated into the optimization model through simulations. The objective of the optimization model is set as a net present value formulation. We also consider priority, budget and labor constraints. Besides, we set hard objectives such as target improvements of different types of costs (e.g., logistics costs).

The remainder of this chapter is organized as follows: Sect. 6.2 provides an overview of the relevant literature on project portfolio optimization and management (PPOM). Section 6.3 describes the proposed model for project portfolio selection. The empirical results are presented in Sect. 6.4. Section 6.5 concludes with a discussion and directions for future research.

6.2 Literature Review

As the knowledge-based economy advanced, project management has been adopted as an organizational form and management technique (Gleadle et al. 2012). In multi-project companies, given limited financial and human resources, it is possible to prioritize and select projects using different optimization criteria. A systematic and rational method for this purpose is portfolio management whose origins date back to 1950s. As Ahmad et al. (2016) stated, portfolio management could help companies in the following ways: (1) to align the portfolio with the company's strategy; (2) to use companies' resources in a better way; (3) to terminate poorly performing or outdated projects; and (4) to monitor key activities constantly to take necessary actions. There exists a vast amount of literature discussing different aspects of project portfolio optimization and management (PPOM). In this section, we provide a review of the related literature focusing both on the contextual and methodological aspects of this topic.

There have been many studies that investigated the implications of different contextual dimensions (i.e. environment, type of business, etc.) on PPOM. Petit and Hobbs (2010) examined the effect of uncertainty on project portfolio management

(PPM) in dynamic environments. Using four portfolios in two large multi-divisional corporations, they identified some types and sources of change in addition to portfolio performance and business strategy changes. IT companies generally execute multi-projects in a dynamic environment as well. Thus, they often make use of PPOM techniques. Vaughan (2011) reported that a good IT project portfolio management can yield reduced costs, better resource allocation, greater agility, increased communication and collaboration into IT investments, and elevation of IT's role within the organization. Drechsler et al. (2010) discussed the role of soft factors such as communication, the project culture and the level of trust among the stakeholders in the failure of IT projects. Inspired from the systemic theory of the German sociologist Niklas Luhmann, the author proposed an approach to identify the true roots of failures by treating projects as social constructs. Software companies—where an effective portfolio management is crucial—mainly operate on the basis of project management. The traditional portfolio management has been adapted for agile and lean software companies in Ahmad et al. (2016). The authors found that agile and lean companies relied on and used tools and methods that solicited customer feedback and engagement. Additionally, Kanban was found to be a significant factor in terms of limiting the number of offerings in the portfolio and their strategic alignment. Regarding the prioritization of iterative (cyclical) projects, Browning and Yassine (2016) pointed out the necessity for a new approach instead of treating such projects as acyclical. They examined several priority rules considering more than 55,000 iterative projects and found that network density, iteration intensity, resource loading profile, and amount of resource contention are the most significant factors that comprises the best priority rule. Korhonen et al. (2016) conducted an in-depth case study in a global OEM in Finland and provided findings related to the management of customization and component selection within multi-project portfolio management. They concluded that there was a two-way interaction between the component–commonality innovation and the value and management of multi-project operations. Brook and Pagnanelli (2014) presented a framework of portfolio management for different types of innovation projects (i.e. breakthrough, platform, and derivative) considering three sustainability dimensions (i.e. ecological, social and economic). After reviewing the current state of the art on decision aid methods, the authors applied their proposed framework of project prioritization and selection in the automotive industry. On the management of innovation, Sicotte et al. (2014) showed that innovation portfolio management can be viewed as a component of dynamic capability using survey data from a sample of more than 900 firms. Yung and Siew (2016) also proposed a way to integrate sustainability thinking into PPM at the screening and the optimal portfolio selection stages. Sustainability project criteria are used at the screening stage to account for uncertainties and then an efficient frontier is developed for selecting an optimal portfolio. Industry 4.0 is envisioned to encompass almost every type of business which may operate in various types of settings. Therefore, the aforementioned dimensions considered in the related literature will greatly help companies in PPM required for the digital transformation of businesses.

As Ellery and Hansen (2012) stated, a project has four dimensions: specification, timeline, resources, and risks. Sanchez et al. (2008) was the first to adapt risk management concepts into PPM to achieve its two most important objectives: portfolio value maximization and project alignment to the organization's strategic goals. Their framework considered resource, knowledge and strategy interrelations. A portfolio is successful if its value is maximized, it is aligned with business strategy, and it is balanced with synergies (Yang and Xu 2017). In an attempt to investigate the relationship between portfolio risk management and portfolio success, Teller (2013) proposed a conceptual framework based on a comprehensive review of the literature and asserted that organization, process, and culture were the most relevant components of portfolio risk management to achieve portfolio success. Teller et al. (2014) added that formal risk management at the project level and integration of risk information at the portfolio level were positively associated with overall project success. Sanchez and Robert (2010) developed key strategic performance indicators to measure the overall strategic performance of portfolios, considering the collection of performances of individual projects and their strategic interdependences. Focused on the objective of strategic alignment in PPM, Kaiser et al. (2015) studied the effects of fundamental strategic changes on project selection and organizational structure through case studies in the German construction industry. They made a significant contribution to the related literature with their substantive theory that integrated strategy implementation, organizational information processing, and structural adaptation. Martinsuo and Killen (2014) explored the evaluation of the strategic value of project portfolio and particularly its non-commercial dimensions. Different from the prior research, they focused on the value that was generated as an outcome of the projects in a firm's portfolio. They emphasized the need for future research to quantify the strategic value in PPM, including ecological, social, health and safety, societal influence, learning and knowledge development, and longer term business value. Pajares and López (2014) identified possible interactions among individual project variables such as schedule, risk and cash flows and showed that they had significant effects on the composition of a portfolio strategically, financially and operationally.

The constituents of portfolio management such as activities (i.e. prioritization, selection, control) and entities (i.e. managers/leaders, stakeholders) play a significant role in the success of a portfolio. Müller et al. (2008) introduced the concept of portfolio control (i.e. portfolio reporting and portfolio selection) into the related literature and analyzed its association with portfolio management performance (i.e. achieving results and purposes) in different contexts (i.e. governance type, industry, geography, dynamics, project type). Their findings indicated that different portfolio control practices were associated with different measures of portfolio management performance and portfolio selection was in positive association with achieving results and achieving purpose. ter Mors et al. (2010) provided a comparison of the theory and the practice in implementing PPM of organizations and reported possible pitfalls organizations might encounter and discussed ways to avoid these. Morris and Geraldi (2011) proposed an additional level of conceptualization which was in institutional level. In earlier research, technical and strategic levels which operated within the project had been identified. Management in the institutional level was concerned with creating conditions to support projects in the parent organization and in the external environment. They claimed that leadership was important at all three levels. To succeed in the management of project portfolios, involvement of stakeholders by providing feedback on the portfolio metrics, criteria, project prioritization, and so on is vital (Vaughan 2011). Beringer et al. (2012) investigated the engagement and behavior of stakeholders in the PPM process using a survey consisting of more than 200 project portfolios from medium to large firms. Koh and Crawford (2012) aimed to identify the theoretical gap in the related literature regarding the role of managers in PPM. They employed an inductive interview-based approach with portfolio managers from service and manufacturing organizations in Australia.

There is a vast amount of research employing different techniques (alternating from approaches the traditional mean-variance technique emanated to those based on computational intelligence) to optimize project portfolios with varying features. Leong and Lim (1991) provided an extension of financial portfolio theory based on the "mean-variance" approach to a more complex domain. They presented a systematic and rigorous way of making resource allocation decisions on project portfolio. Bouri et al. (2002) illustrated the use of multi-criteria approach in portfolio selection process. They criticized the traditional methods of portfolio optimization (i.e. mean-variance criterion) which was based only on an economic function. They rather suggested using a set of criteria which were more representative of reality. Similarly, Smith et al. (2012) proposed a methodology for portfolio selection in the presence of multiple stakeholders and complex constraints using multi-objective optimization. Lin et al. (2005) employed fuzzy set theory and portfolio matrices for the overall assessment of portfolio competitiveness. They formulated a binary integer linear programming model to allocate constrained resources and select the optimal portfolio incorporating the diversity of confidence and optimism levels of decision makers. Rajapakse et al. (2006) proposed a method to compute the value of different portfolios for drug companies incorporating the uncertainties in the development phase. Ellery and Hansen (2012) described PPM in this context as "the science and practice of determining the optimal constellation of projects to support a drug company in achieving its strategic goals." Based on survey data, Rosacker and Olson (2008) aimed to identify and describe the project selection and evaluation techniques in IT project management, which is concerned with the improvement of returns on IT investments and the overall alignment of IT with the business (Vaughan 2011). Using an artificial neural network, Constantino et al. (2015) described the design, development and testing stages of a decision support system to classify the level of project riskiness considering a set of critical success factors and the experience of project managers. Inquiring to what extent a firm can diversify away its total operational risk given the efficiency and riskiness of projects and the existence of positively cross-correlated projects, Paquin et al. (2015) developed a probabilistic model based on net present value for project portfolio risk diversification. They assessed the effectiveness of representative capital investment projects added to a firm's project portfolio in reducing its total operational risk. El Hannach et al. (2016) aimed to assess the inaccuracy of evaluators' judgements when employing a multi-criteria decision making approach in PPM. Through a case study, they applied the information entropy method for prioritizing and selecting an appropriate project portfolio management information system. Brito et al. (2017) extended the mean-variance criterion of Markowitz (1952) using higher moments (e.g. skewness and kurtosis) within a bi-objective model. They analyzed the trade-off between expected utility and cardinality. Developing a decision support model, Yang and Xu (2017) investigated the effect of portfolio management practices on new product development performance in multi-project environments under uncertainty and found that top management involvement, manager competency, portfolio management process design and implementation, and project termination were the most important practices to firms.

Industry 4.0 transformation will entail multi-staged and comprehensive portfolio management practices so that interrelated projects can be carried out simultaneously taking their interdependencies into account. Under such interrelations and high uncertainties, the optimization of such portfolios using traditional approaches stemmed from the mean-variance technique cannot yield reliable results. This study proposes simulation-based optimization as an alternative approach to be employed in multi-dimensional environments in the presence of high uncertainty. In this chapter, we develop a simulation-based optimization model with the aim of maximizing the total profit of project portfolios in the framework of Industry 4.0 transformation.

6.3   Project Portfolio Optimization Model

Since determining optimal project portfolio for digital transformation among various project alternatives requires the consideration of multiple constraints and interdependencies, an integer programming model is proposed to address this problem. In the context of linear programming (LP), the objective function should be a function of the decision variables. An LP will either minimize or maximize the value of the objective function. Finally, the decisions that must be made are subject to certain requirements and restrictions of a system which are called constraints. The following assumptions were made in the model:

1. Project alternatives are known.
2. Project investment budget is known and fixed.
3. Savings for each project is stochastic.
4. Some relationships (mutually exclusiveness, precedence, either-or) among projects exist.
5. Some of the projects are mandatory.

The formulation and detailed information about the optimization model are presented below.

Indices

i projects
m mutually exclusiveness relation
w either-or relation

Parameters

B	Budget for project portfolio
I	Number of projects
IC_i	Investment cost for project i
$ES_i(\mu, \sigma)$	Net present value of energy savings for project i with a mean μ and standard deviation σ
$LS_i(\mu, \sigma)$	Net present value of labor savings for project i with a mean μ and standard deviation σ
$MS_i(\mu, \sigma)$	Net present value of material savings for project I with a mean μ and standard deviation σ
TES	Threshold value for portfolio energy saving
TLS	Threshold value for portfolio labor saving
TMS	Threshold value for portfolio material saving
PR_i	The set of predecessors of project i
M	The number of exclusiveness relations
ME_m	The set of projects for mutually exclusiveness relation m
W	The number of either-or relations
EO_w	The set of projects for either-or relation w
MP	The set of mandatory projects

Variables

$$x_i = \begin{cases} 1; \text{if project } i \text{ is selected in the portfolio} \\ 0; \text{otherwise} \end{cases} i = 1, \dots, I \qquad (6.1)$$

Objective function

$$\text{Maximize } Z = \sum_{i=1}^{I} (ES_i(\mu, \sigma) + LS_i(\mu, \sigma) + MS_i(\mu, \sigma) - IC_i) * x_i \qquad (6.2)$$

subject to

$$\sum_{i=1}^{I} IC_i * x_i \leq B \quad i = 1, \dots, I \qquad (6.3)$$

$$\sum_{i=1}^{I} ES_i(\mu, \sigma) * x_i \geq TES \qquad (6.4a)$$

$$\sum_{i=1}^{I} LS_i(\mu, \sigma) * x_i \geq TLS \qquad (6.4b)$$

$$\sum_{i=1}^{I} MS_i(\mu, \sigma) * x_i \geq TMS \qquad (6.4c)$$

$$x_f - x_i \geq 0 \quad i = 1, \ldots, I \quad \forall f : f \in PR_i \qquad (6.5)$$

$$\sum_{i \in ME_m} x_i \leq 1 \quad m = 1, \ldots, M \qquad (6.6)$$

$$\sum_{i \in EO_w} x_i = 1 \quad w = 1, \ldots, W \qquad (6.7)$$

$$x_i = 1 \quad \forall i : i \in MP \qquad (6.8)$$

$$x_i \in \{0, 1\} \quad i = 1, \ldots, I \qquad (6.9)$$

The objective function (Eq. 6.2) maximizes the total profit (total savings-total investment cost) of the project portfolio. Constraint (Eq. 6.3) implies that portfolio budget cannot be exceeded. Constraints (Eq. 6.3) satisfy that the portfolio provides energy, labor and material savings with respect to pre-determined threshold values.

Regarding dependencies among projects, constraints (Eq. 6.5) ensure the precedence relations among projects. Constraints (Eq. 6.6) allow that at most one of the project is included to the portfolio for each mutually exclusiveness relation set. Constraints (Eq. 6.7) verify that exactly one project is selected for each either-or relation. Constraints (Eq. 6.8) provide the inclusion of mandatory projects. Constraints (Eq. 6.9) define the domains for variables.

6.4   Application

The model was applied to the hypothetical project alternatives of an automotive manufacturer over a 5-year planning horizon. The detail of the projects is given in Table 6.1. Note that the monetary unit is dollars ($) and the project savings are normally distributed. The formulated problem was solved via Crystal Ball Version 7.2.1. Additionally, the model is simulated for 1000 runs and 1000 trials are executed in each simulation run to minimize the possible errors arising from the random variables. Portfolio budget is regarded as $750,000 for the 5-year planning period. Threshold values for energy, labor and material savings are $140,000, $220,000, $60,000, respectively.

The application is formulated as follows.

Table 6.1 The detail of projects

No	Project	Investment cost ($)	Energy saving parameters (μ, σ)	Labor saving parameters (μ, σ)	Material saving parameters (μ, σ)	Statement
1	IOT for energy consumption	90,000	(120,000, 12,000)	(10,000, 2000)	(12,000, 1200)	Preceded by project 5
2	AR in logistics (Order Picking System)pick by vision	100,000	(50,000, 5000)	(70,000, 14,000)	(10,000, 1000)	Mutually exclusive 3, 4
3	AR in maintenance	80,000	(70,000, 7000)	(10,000, 2000)	(10,000, 1000)	Mutually exclusive 2, 4
4	AR in process control	65,000	(25,000, 2500)	(40,000, 8000)	(30,000, 3000)	Mutually exclusive 2, 3
5	Manufacturing process traceability (Sensor/RFID)	120,000	(15,000, 1500)	(90,000, 18,000)	(60,000, 6000)	Mandatory
6	Data analytics project for predictive maintenance	25,000	(30,000, 3000)	(6000, 1200)	(5000, 500)	Preceded by project 5
7	Virtual simulation system	35,000	(15,000, 1500)	(25,000, 5000)	(7000, 700)	Preceded by project 5
8	Robotics for packaging	60,000	(15,000, 1500)	(55,000, 11,000)	(6000, 600)	Independent
9	Additive manufacturing for experimental prototypes	90,000	(15,000, 1500)	(20,000, 4000)	(80,000, 8000)	Independent
10	Indoor localization for moving assets	55,000	(30,000, 3000)	(10,000, 2000)	(35,000, 3500)	Preceded by project 5
11	Robotics for kitting	65,000	(25,000, 2500)	(45,000, 9000)	(10,000, 1000)	Either or with project 13
12	Robotics for test and inspection	75,000	(14,000, 1400)	(54,000, 10,800)	(22,000, 2200)	Independent
13	Robotics for material loading and unloading	70,000	(45,000, 4500)	(50,000, 10,000)	(5000, 500)	Either or with project 11

(continued)

Table 6.1 (continued)

No	Project	Investment cost ($)	Energy saving parameters (μ, σ)	Labor saving parameters (μ, σ)	Material saving parameters (μ, σ)	Statement
14	Additive manufacturing for tool design	55,000	(20,000, 2000)	(25,000, 5000)	(40,000, 4000)	Independent
15	Additive manufacturing highly customized product	60,000	(19,000, 1900)	(21,000, 4200)	(32,000, 3200)	Independent
16	Analytics engine integration in the product cloud platform	55,000	(0, 0)	(72,000, 14,400)	(0, 0)	Mandatory

Maximize Z $=[ES_1(120000, 12000) + LS_1(10000, 2000) + MS_1(12000, 1200) - 90000] * x_1 + [ES_2(50000, 5000)$

$\quad + LS_2(70000, 14000) + MS_2(10000, 1000) - 100000] * x_2 + [ES_3(70000, 7000) + LS_3(10000, 2000)$

$\quad + LS_2(70000, 14000) + MS_2(10000, 1000) - 100000] * x_2 + [ES_3(70000, 7000) + LS_3(10000, 2000)$

$\quad + [ES_5(15000, 1500) + LS_5(90000, 18000) + MS_5(60000, 600) - 120000] * x_5$

$\quad + [ES_6(30000, 3000) + LS_6(6000, 1200)$

$\quad + MS_6(5000, 500) - 25000] * x_6 + \big[ES_7(15000, 1500) + LS_7(25000, 5000)$

$\quad + MS_7(7000, 700) - 35000] * x_7$

$\quad + [ES_8(15000, 1500) + LS_8(55000, 11000) + MS_8(6000, 600) - 60000] * x_8$

$\quad + [ES_9(15000, 1500) + LS_9(20000, 4000)$

$\quad + MS_9(80000, 8000) - 90000] * x_9 + [ES_{10}(30000, 3000) + LS_{10}(10000, 2000)$

$\quad + MS_{10}(35000, 3500) - 55000] * x_{10}$

$\quad + [ES_{11}(25000, 2500) + LS_{11}(45000, 9000) + MS_{11}(10000, 1000) - 65000] * x_{11}$

$\quad + [ES_{12}(14000, 1400)$

$\quad + LS_{12}(54000, 10800) + MS_{12}(22000, 2200) - 75000] * x_{12}$

$\quad + [ES_{13}(45000, 4500) + LS_{13}(50000, 10000)$

$\quad + MS_{13}(5000, 500) - 70000] * x_{13}$

$\quad + [ES_{14}(20000, 2000) + LS_{14}(25000, 5000) + MS_{14}(40000, 4000) - 55000] * x_{14}$

$\quad + [ES_{15}(19000, 1900) + LS_{15}(21000, 4200) + MS_{15}(32000, 3200) - 60000] * x_{15}$

$\quad + [ES_{16}(0, 00) + LS_{16}(72000, 14400)$

$\quad + MS_{16}(0, 0) - 55000] * x_{16}$

subject to

$90000 * x_1 + 100000 * x_2 + 80000 * x_3 + 65000 * x_4 + 120000 * x_5 + 25000 * x_6 + 35000 * x_7$

$\quad + 60000 * x_8 + 90000 * x_9 + 55000 * x_{10} + 65000 * x_{11} + 75000 * x_{12} + 70000 * x_{13} + 55000 * x_{14}$

$\quad + 60000 * x_{15} + 55000 * x_{16} \leq 750000$

$ES_1(120000, 1200) * x_1 + ES_2(50000, 5000) * x_2 + ES_3(70000, 7000) * x_3 + ES_4(25000, 2500) * x_4$

$\quad + ES_5(15000, 1500) * x_5 + ES_6(30000, 3000) * x_6 + ES_7(15000, 1500) * x_7 + ES_8(15000, 1500 * x_8$

$\quad + ES_9(15000, 1500) * x_9 + ES_{10}(30000, 3000) * x_{10} + ES_{11}(25000, 2500) * x_{11}$

$\quad + ES_{12}(14000, 1400) * x_{12} + ES_{13}(45000, 4500) * x_{13}$

$\quad + ES_{14}(20000, 2000) * x_{14} + ES_{15}(19000, 1900) * x_{15} + ES_{16}(0, 0) * x_{16} \geq 140000$

$LS_1(10000, 2000) * x_1 + LS_2(70000, 14000) * x_2 + LS_3(10000, 2000) * x_3 + LS_4(40000, 8000) * x_4$

$\quad + LS_5(90000, 18000) * x_5 + LS_6(6000, 1200) * x_6 + LS_7(25000, 5000) * x_7$

$\quad + LS_8(55000, 1100 * x_8 + LS_9(20000, 4000) * x_9 + LS_{10}(10000, 2000) * x_{10}$

$\quad + LS_{11}(45000, 9000) * x_{11} + LS_{12} + (154000, 10800) * x_{12} + LS_{13}(50000, 10000) * x_{13}$

$\quad + LS_{14}(25000, 5000) * x_{14} + LS_{15}(21000, 4200) * x_{15} + LS_{16}(72000, 14400) * x_{16} \geq 220000$

$MS_1(12000, 1200) * x_1 + MS_2(10000, 1000) * x_2 + MS_3(10000, 1000) * x_3 + MS_4(30000, 3000) * x_4$

$\quad + MS_5(60000, 6000) * x_5 + MS_6(5000, 500) * x_6 + MS_7(7000, 700) * x_7 + MS_8(6000, 600 * x_8$

$\quad + MS_9(80000, 8000) * x_9 + MS_{10}(35000, 3500) * x_{10} + MS_{11}(10000, 1000) * x_{11}$

$\quad + MS_{12}(22000, 2200) * x_{12} + MS_{13}(5000, 500) * x_{13}MS_{14}(40000, 4000) * x_{14}$

$\quad + MS_{15}(32000, 3200) * x_{15} + MS_{16}(0, 0) * x_{16} \geq 60000$

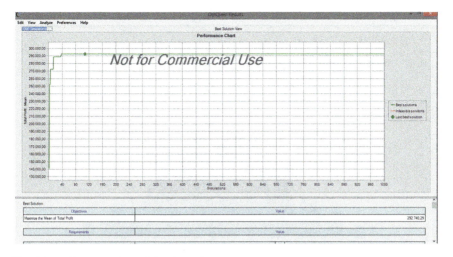

Fig. 6.1 Performance chart of the simulation

$$x_5 - x_1 \geq 1$$
$$x_5 - x_6 \geq 1$$
$$x_5 - x_7 \geq 1$$
$$x_5 - x_{10} \geq 1$$

$$x_2 + x_3 + x_4 \leq 1$$

$$x_{11} + x_{13} = 1$$

$$x_{16} = 1$$
$$x_5 = 1$$

$$x_1, x_2, x_3, x_4, x_5, x_6, x_7, x_8, x_9, x_{10}, x_{11}, x_{12}, x_{13}, x_{14}, x_{15}, x_{16} \in \{0, 1\}$$

The results showed that eleven projects (Projects 1, 4, 5, 6, 7, 8, 9, 10, 13, 14 and 16) were selected to be planned for project portfolio. The analysis of the results further shows that the total of 11 projects selected in the period accounts for 96% of the total budget. The simulation yielded an objective function value with a mean value of $292,740.29 and standard deviation value of $35,920.87. The performance chart of the simulation is given in Fig. 6.1.

According to Fig. 6.2, the total profit varies between $174,968.43 and $389,965.19. Furthermore, the data of the total benefit fit to Student's t-distribution.

As shown in Table 6.2, the expected total profit in a five-year-horizon is greater than or equal to $247,715.44; $275,133.58; $293,095.75 and $311,651.19 with probabilities 0.9, 0.7, 0.5, and 0.3, respectively.

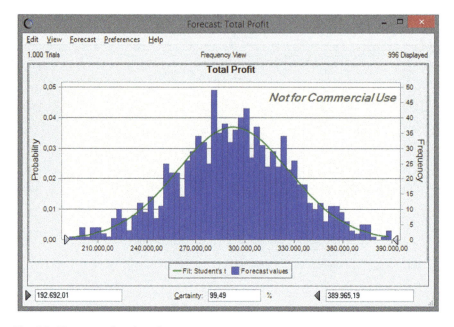

Fig. 6.2 Histogram of total profit

Table 6.2 Lower limits of the total profit for different probabilities

Probability	Total profit value ($)
%100	≥ 174,968.43
%90	≥ 247,715.44
%80	≥ 264,537.86
%70	≥ 275,133.58
%60	≥ 284,050.57
%50	≥ 293,095.75
%40	≥ 301,577.10
%30	≥ 311,651.19
%20	≥ 323,422.81
%10	≥ 337,629.71
%0	≥ 389,965.19

6.5   Conclusion

The processes associated with project portfolio management bring the operations and projects functions together to fulfil and maximize the contribution of projects to the overall welfare and success of the enterprise. Executives expect projects to be aligned with strategies and make effective use of the limited resources, and deliver certain benefits. In this context, determining the right mix of projects for the portfolio is considered as one of the most important components for PPM. In order to manage project portfolios, organizations must utilize methods of prioritizing and selecting projects from a project pool with multiple constraints and interdependencies.

Due to the high investment costs and high level of technological variety in Industry 4.0 projects, the economic achievement is provided through digital transformation on the condition that project portfolio selection is performed precisely, especially for increasing number of projects with interrelations. In this chapter, an optimization model is presented to address this problem. Since it is generally difficult to determine the exact values for project savings in real life applications, the main contribution of this model is that energy, material and labor savings are considered to be stochastic.

To demonstrate its effectiveness and practicality, the proposed optimization model is applied to the Industry 4.0 project alternatives of an automotive manufacturer. Since the model is reflected on a spreadsheet and the solution as well as the analysis are executed via simulation based optimization, working environment of the model provides convenience for quickly quantifying the project portfolio and adapting to the changes in it.

As an avenue for future research, project risk factors can be integrated to the proposed model. Moreover, project alternatives may be grouped with respect to digital transformation phases. Another research direction would be to consider this problem as multi-objective.

References

Ahmad MO, Lwakatare LE, Kuvaja P, Oivo M, Markkula J (2016) An empirical study of portfolio management and Kanban in agile and lean software companies. J Software: Evol Process 1–16

Beringer C, Jonas D, Georg Gemünden H (2012) Establishing project portfolio management: an exploratory analysis of the influence of internal stakeholders' interactions. Project Management Journal 43(6):16–32

Bouri A, Martel JM, Chabchoub H (2002) A multi-criterion approach for selecting attractive portfolio. Journal of Multi-Criteria Decision Analysis 11(4–5):269–277

Brito RP, Sebastião H, Godinho P (2017) Portfolio management with higher moments: the cardinality impact. Int Trans Operational Res, [e-journal], http://dx.doi.org/10.1111/itor.12404

Brook JW, Pagnanelli F (2014) Integrating sustainability into innovation project portfolio management–a strategic perspective. J Eng Tech Manage 34:46–62

Browning TR, Yassine AA (2016) Managing a portfolio of product development projects under resource constraints. Decis Sci 47(2):333–372

Costantino F, Di Gravio G, Nonino F (2015) Project selection in project portfolio management: an artificial neural network model based on critical success factors. Int J Project Manage 33 (8):1744–1754

De Reyck B, Grushka-Cockayne Y, Lockett M, Calderini SR, Moura M, Sloper A (2005) The impact of project portfolio management on information technology projects. Int J Project Manage 23(7):524–537

Drechsler A, Kalvelage P, Trepper T (2010) Systemic IT project management: a rational way to manage irrationalities in IT projects? In: Harmsen et al (ed) Working conference on practice-driven research on enterprise transformation The Netherlands, 11 November 2010. Springer Berlin Heidelberg, Germany, pp 127–155

El Hannach D, Marghoubi R, Dahchour M (2016) Project portfolio management information systems (PPMIS): information entropy based approch to prioritize PPMIS. In: 4th IEEE international colloquium on information science and technology (CiSt'16). Morocco, 24–26 October 2016, pp 228–234

Ellery T, Hansen N (2012) Pharmaceutical lifecycle management: making the most of each and every brand. Wiley, Hoboken

Gleadle P, Hodgson D, Storey J (2012) 'The ground beneath my feet': projects, project management and the intensified control of R&D engineers. New Technol, Work Employment 27(3):163–177

Jeffery M, Leliveld I (2004) Best practices in IT portfolio management. MIT Sloan Manage Rev 45(3):41–49

Kaiser Michael G, El Arbi Fedi, Ahlemann Frederik (2015) Successful project portfolio management beyond project selection techniques: understanding the role of structural alignment. Int J Project Manage 33(1):126–139

Koh A, Crawford L (2012) Portfolio management: the Australian experience. Proj Manage J 43 (6):33–42

Korhonen T, Laine T, Lyly-Yrjänäinen J, Suomala P (2016) Innovation for multiproject management: the case of component commonality. Proj Manage J 47(2):130–143

Leong SM, Lim KG (1991) Extending financial portfolio theory for product management. Decis Sci 22(1):181–193

Lin C, Tan B, Hsieh PJ (2005) Application of the fuzzy weighted average in strategic portfolio management. Decis Sci 36(3):489–511

Markowitz H (1952) Portfolio selection. J Finan 7(1):77–91

Martinsuo M, Killen CP (2014) Value management in project portfolios: identifying and assessing strategic value. Proj Manage J 45(5):56–70

Morris PW, Geraldi J (2011) Managing the institutional context for projects. Proj Manage J8 42 (6):20–32

Müller R, Martinsuo M, Blomquist T (2008) Project portfolio control and portfolio management performance in different contexts. Proj manage j 39(3):28–42

Pajares J, López A (2014) New methodological approaches to project portfolio management: the role of interactions within projects and portfolios. Procedia-Social Behavioral Sci 119:645–652

Paquin JP, Tessier D, Gauthier C (2015) The effectiveness of portfolio risk diversification: an additive approach by project replication. Proj Manage J 46(5):94–110

Petit Y, Hobbs B (2010) Project portfolios in dynamic environments: sources of uncertainty and sensing mechanisms. Project Management Journal 41(4):46–58

Pinto JK (2016) Project management: achieving competitive advantage, 4th edn. Pearson Education Inc, Malaysia

Rajapakse A, Titchener-Hooker NJ, Farid SS (2006) Integrated approach to improving the value potential of biopharmaceutical R&D portfolios while mitigating risk. J Chem Technol Biotechnol 81(10):1705–1714

Rosacker KM, Olson DL (2008) An empirical assessment of IT project selection and evaluation methods in state government. Proj Manage J 39(1):49–58

Sanchez H, Robert B (2010) Measuring portfolio strategic performance using key performance indicators. Proj Manage J 41(5):64–73

Sanchez H, Robert B, Pellerin R (2008) A project portfolio risk-opportunity identification framework. Proj Manage J 39(3):97–109

Sicotte H, Drouin N, Delerue H (2014) Innovation portfolio management as a subset of dynamic capabilities: measurement and impact on innovative performance. Proj Manage J 45(6):58–72

Smith P, Ferringer M, Kelly R, Min I (2012) Budget-constrained portfolio trades using multiobjective optimization. Syst Eng 15(4):461–470

Teller J (2013) Portfolio risk management and its contribution to project portfolio success: an investigation of organization, process, and culture. Proj Manage J 44(2):36–51

Teller J, Kock A, Gemünden HG (2014) Risk management in project portfolios is more than managing project risks: a contingency perspective on risk management. Proj Manage J 45 (4):67–80

ter Mors M, Drost R, Harmsen F (2010) Project portfolio management in practice. In: Harmsen et al (ed) Working conference on practice-driven research on enterprise transformation, The Netherlands, 11 November 2010. Springer Berlin Heidelberg, Germany, pp 107–126

Vaughan P (2011) Information technology project management. In: Lane D (ed) The Chief Information Officer's Body of Knowledge: people, Process, and Technology, 1st edn. Wiley, Hoboken, pp 123–128

Yang Y, Xu DL (2017) A methodology for assessing the effect of portfolio management on NPD performance based on Bayesian network scenarios. Expert Syst, [e-journal] 34(2). https://doi.org/10.1111/exsy.12190

Yung R, Siew J (2016) Integrating sustainability into construction project portfolio management. KSCE J Civ Eng 20(1):101–108

Chapter 7
Talent Development for Industry 4.0

Gaye Karacay

Abstract In today's global environment, sustainability and competitive advantage of companies depend mostly on their capability of adaptation to changing business requirements. The Fourth Industrial Revolution, driving from the advancements in new digital technologies known collectively as Industry 4.0, has been profoundly changing dynamics of most industries. Automation of business processes together with emergence of novel business models impose new digital skill requirements for workforce. Creating future workforce involves not only attracting and developing new talent needed, but also re-skilling current employees through training programs as well as re-designing work processes for reducing the skill mismatch between jobs and employees. This chapter examines how Industry 4.0 will alter the landscape for talent development.

7.1 Introduction

We are currently witnessing to the Fourth Industrial Revolution, that is, the new era of industrialization derived from innovative forms of decentralized as well as complex production processes. *Industry 4.0* has been used as a common description for these novel production processes which are completely automated via technology, and devices communicating autonomously with each other along the value chain activities. Real time data exchange between machines and materials, gradually more autonomous production systems, and additive manufacturing techniques has been profoundly transforming the dynamics of most industries. These transformations bring about *creation of new sectors* like data science, *creation of new business models* such as platforms*, creation of new types of companies* such as cloud computing providers; and *initiation of new organizational roles* like social media account managers. By interacting with other socio-economic and demographic factors,

G. Karacay (✉)
Department of Industrial Engineering, Faculty of Management,
Istanbul Technical University, 34367 Maçka, Istanbul, Turkey
e-mail: karacayaydin@itu.edu.tr

© Springer International Publishing Switzerland 2018
A. Ustundag and E. Cevikcan, *Industry 4.0: Managing The Digital Transformation*,
Springer Series in Advanced Manufacturing, https://doi.org/10.1007/978-3-319-57870-5_7

significant changes in business models activate major disruptions in labor markets. Ongoing shifts towards knowledge based economies necessitate development of new digital skill sets for workforce which would eventually transform how and where people work (Beechler and Woodward 2009; Guthridge et al. 2008).

Developments in robotics, artificial intelligence, and machine learning are channeling through a new phase of automation of work processes at which 'smart' and even 'smarter' machines match or even outperform human performance in a wide range of activities. According to a recent report by McKinsey Global Institute (2017) derived from evaluations of 2000 work activities across 800 occupations, by adopting only the well-established technologies almost 5% of all occupations can be fully automated while 60% of all occupations have at least 30% of modules that could be automated by the current technology on hand. Automation certainly raise total productivity even mostly outperform human performance by reducing human error, improving quality, and maximizing speed of processes. Nevertheless, the pace and extent of automation of all work activities and occupations will not be quite similar due to the differences in the technical feasibility, potential economic gain, and probable social cost of technology adoptions required within and across various industries.

While all occupations are made up of activities with different potential for automation, the work activities that have highest potential for immediate automation typically contain tasks that depend on pre-specified routine physical activities, data collection and processing (Frey and Osborne 2013). Evidenced by the past experiences, the impact of technological innovations and automation on labor markets mostly indicate decline of employment in routine intensive occupations—i.e. occupations consist of tasks which follow a detailed and specific course of actions so that they could be easily performed by complex algorithms (Frey and Osborne 2013; Autor et al. 2003). McKinsey Global Institute Report (2017) points that occupations of transportation, office administration, production, and food preparation have relatively higher potential for technical automation due to their activity sets being primarily based on predictable physical activities as well as data administration. On the other hand, occupations such as management, personal care, and sales whose activity sets mostly involve managing and developing people, applying expertise to decision making, planning, and creative tasks, interfacing with stakeholders, performing physical activities with operating machinery in unpredictable environments have lower potential for technical automation.

An earlier study by Frey and Osborne (2013) found similar findings, except for service occupations. According to Frey and Osborne (2013), service occupations are also at high risk of being automated evident by the rising usage of service robots in various sectors. That said, they further specified that human workers still have a relative advantage in tasks requiring creativity, perception, and social intelligence. Likewise, Brynjolfsson and McAfee (2011) argue that as a result of advanced pattern recognition abilities by means of sophisticated algorithms on big data, non-routine cognitive tasks have also become feasible for automation. They name Google's driverless car and IBM's Watson as proofs for how far and how fast computers are progressing in pattern recognition and complex communication abilities which have been thought to be exclusively human capabilities. Different combinations of digital technologies are enabling creation of machine intelligence by which computers have started to perform cognitive tasks such as language and voice recognition that they could never have done previously. Recent developments in computing, robotics and artificial intelligence enable automation and digital systems to penetrate the group of tasks that until recently used to be done solely by human power by means of cognitive capabilities such as sensing, reasoning, and decision making. Brynjolfsson and McAfee (2011) argue that computers' cognitive capabilities are growing exponentially even faster than *Moore's law*, i.e. an allegation by Intel microprocessor architect Gordon Moore in 1965 regarding the doubling number of transistors in a minimum-cost integrated circuit every 12 months, and continuation of the same rate of improvement in future. The rapid and significant developments in computer skill and abilities not only make automation more feasible to substitute human workforce in a wider range of occupations, but also change the nature and scope of work across industries and occupations. As machine intelligence capacities develop further, applications of artificial intelligence like machine learning, 3D printers, driverless cars, and the others are likely to abolish more jobs currently done by humans, not just in manufacturing but also in service industries, ranging from low-skill tasks like home delivery to high-skill professional tasks like buying and selling stocks in the stock markets. While some researchers and experts declare that automation will eventually replace human workforce to a large extent, some others claim that it is not possible to massively substitute human workforce by means of automation, thus digital systems would only be used for assisting human workforce even in more sophisticated digital platforms (Autor 2015; Autor and Handel 2013; Frey and Osborne 2013).

On-going advancements in sensor and radio-frequency identification (RFID) technologies make real-time data tracking ever more easier, and facilitate exponential growth of "Internet of Things". It is projected that by 2020 nearly 50 billion items will be connected to each other (Cisco 2011). Consequently, more and more systems can be much efficiently digitalized. Within this context, sooner or later more occupations will be automated away; but at the same time new jobs and occupations will emerge as well as hybrid configurations will be formed through human-machine integrations. As digital transformations increase and become much more feasible, they would bring about new business models. By interacting with other socio-economic factors, significant changes in business models trigger disruptions in labor markets by creating new job descriptions and new job roles. Some possible new job roles of the future include; robot coordinator, digital product manager, digital business developer, data protection officer, web project manager, web integrator, digital communications planner, digital copywriter, user experience designer, crowd innovation facilitator, social media manager, content curator, digital work experience expert, and design learning manager etc. (e.g., World Economic Forum 2016b). Therefore, although recent advancements in digital technologies may bring about some potential job displacements and job losses mostly in the industries where automation can easily displace tasks and activities traditionally performed by humans, yet the rise of digitalization would also have considerable positive impact on employment by creating new jobs and roles in various industries.

As technology improves and automation becomes more feasible to replace human workforce, workers who would lose their jobs to digital systems need to be moved to tasks that are not at risk of being automated—i.e., tasks requiring hi-tech and social intelligence. The studies by Autor and colleagues (2003) and Berger and Frey (2016) showed that workers who had occupations that previously involved routine tasks progressively performed more analytic and interactive work once their industries experienced fast digitalization. Extensive automation of routine activities will reduce the demand for lower-skill and routine-intensive occupations, while it will increase the need for high-skill workers with novel skills. By increased automation, workers would need to focus more on activities that machines cannot easily accomplish, such as the ones that require cognitive abilities.

Industry 4.0 has been bringing about transformational and massive changes in all layers of industry structures, including the one which has started barely in employment landscape. The shift in workforce dynamics will be tremendous when the adjustment completely occurs. Fulfilling the need of developing future work-force brings some requirements to companies, business leaders and governments. While current employees need to be re-skilled for the requirements of digital economy, prospective employees who are today's youngest generations, need to be educated in line with the requirements of future jobs and skills. In order to get prepared for these imminent changes, and ensure their productivity and competi-tiveness within Industry 4.0 era, organizations need to develop their future work-force while adopting new business models and organizational structures. Creating future workforce involves not only attracting, recruiting and developing new talent needed, but also re-skilling current employees through training programs as well as re-designing work processes to reduce the skill mismatch between jobs and employees. This chapter examines talent development within Industry 4.0, which has emerged as one of the most urgent necessities for organizations in line with recent transformational technological advancements.

7.2   Skill Requirements in the Digital World

As game-changing technologies continue to evolve and further develop, adoption of technology for digital systems would become less expensive and more feasible for wider areas. As a result, there would be more opportunities for computer-based automation and robotic systems to replace as well as complement human workers both in production and service industries. In order to drive the transformational opportunities promised by Industry 4.0 and create value from automation, yet organizations need to consider developing their future workforce with competencies aligned to industry-specific requirements. Rapid and extensive automation of business processes together with emergence of novel business models impose new skill requirements for workforce. Indeed, further adoption of digital systems along with successful implementation of Industry 4.0 require even wider range of employee skills due to increased complexity of work environments with new operational and organizational structures. Consequently, roles of employees will change in terms of content and work processes, and these changes would require significant transformations in jobs and skill profiles of employees.

Future of Jobs Report revealed that by 2020 more than one-third of the desired skill set of most jobs will be comprised of skills which are not yet seen as important today (World Economic Forum 2016a). More surprisingly, 65% of children today will do jobs that haven't even been developed yet (OECD 2016a). According to 2020 predictions of World Economic Forum (WEF 2016a), future workforce is expected to have mostly *cognitive abilities* (52%), *systems skills* (42%), and *complex problem solving skills* (40%). In addition to those skills, workers are required to have the basic *skills for information and communication technologies* (ICT). Corresponding to increased automation and digitalization of work processes, organizations more depend on employees with ICT specializations who can analyze Big Data, make coding, develop applications, and manage complex database networks. Indeed, ICT skills do not only support the infrastructure that firms rely on for their business, but also enable innovation in digital economy to flourish (OECD 2016b; Quintini 2014). ICT skills become a necessity for all employees even for workers having low-skilled jobs. For example, at restaurants waiters need to take orders on iPads, or blue-collar workers have to work in smart factories together with automated systems; and thus, these employees need basic ICT skills to be able to do their jobs in this novel hybrid business structures.

ICT and other 'hard' skills required for Industry 4.0 have a more interdisciplinary character than basic literacy and professional know-how. Employees need to combine technical knowledge of a specific job with a collection of ICT skills varying from basic ICT literacy to advanced programming capabilities. However, execution of these more interdisciplinary hard skills require capabilities of collaboration, communication, and adaptability; in other words, within today's constantly changing work environment 'soft' skills have become even more crucial ever than before. For that reason, in order to perform effectively and sufficiently within Industry 4.0 work circumstances, employees should get into the habit of continuous learning, not simply in their own professions but in a wider angle with an interdisciplinary perspective.

While organizations need to invest in skill sets of their future workforce to be able to utilize the benefits of technological advancements, current workers should develop their technological capabilities as well as acquire new skills for staying in the job market. In this respect, having the capability of adaptation to constant technological advancements is vital both for employees and organizations. Industry 4.0 work systems evidently necessitate employees having degrees in fields related to science, technology, engineering, and mathematics (STEM) so that these employees would have core skills built on these basic sciences required for technology based innovations. STEM competencies are defined as the set of cognitive skills, knowledge, and abilities associated with STEM occupations, and they have become critical for economic competitiveness due to their positive influence on innovation, technological growth, and economic development. According to a report from the Georgetown University Center on Education and the Workforce, latest developments in technology based innovations have led to demand for STEM competencies even beyond traditional STEM occupations (Carnevale et al. 2011). Nowadays, STEM competencies have been required for occupations especially in fast-growing industries, such as healthcare and business services. While the overall employment levels of low-skilled workers in industries like manufacturing, utilities and transportation decrease due to increased automation and digitalization of work processes, high-skilled workers in these industries need to develop STEM competencies to secure their jobs in future. Therefore, STEM education, from high school to postgraduate level, will be necessary for cultivating talent pipelines needed for jobs which utilize STEM competencies to boost innovative capacities.

As automation and digitalization of work processes increase, workers will be required to take charge of less automatable and more complex tasks, whose completion necessitate solid literacy, numeracy, problem-solving, and ICT skills together with soft skills of autonomy, co-ordination and collaboration (Grundke et al. 2017). Overall, there will be higher demand placed on all members of the workforce in terms of managing complexity, problem-solving and higher levels of abstraction for obtaining simplified representation of the bigger wholes. Moreover, there will be a need to co-ordinate between virtual and real machines as well as between manual and robotic systems, hence employees will be expected to act more on their own initiative, have excellent communication skills and be able to organize their own work. As a result, Industry 4.0 environment would allow employees to have more opportunities for individual responsibility, decentralized leadership, and involvement in decision making, all of which triggering a shift in organization of work tasks from Taylorist approaches to more holistic and socio-technical methods that embrace the interaction between people and technology. As further shifts in organizational structures continue to occur from a command-and-control to a more fluid ones, interpersonal and communication skills together with other social skills like emotional intelligence and persuasion will be in much higher demand as a future workforce skill. With the emergence of networked corporate systems, interactions with customers, suppliers, and other stakeholders have become much more dispersed, as a result aligning different strategic goals and building consensus among various parties have become harder than before. Besides, these networked corporate systems bring together different organizations from various geographies, not always be physically but yet digitally, accordingly cultural intelligence, the capability of adapting to and understanding of different cultural requirements, emerges as an important skill. In view of that, in today's digital world hard-skills as technical know-how and specific profession expertise necessitate being supplemented with social skills.

In today's business context, while innovation at its core still requires a solid foundation of knowledge in science, technology, engineering, and mathematics, obtaining economic value from any innovation necessitates developing applications and usage systems which are customized to meet critical individual and social needs (Hill 2007). For instance, Apple's iPod is a good example for understanding these new trends for obtaining economic value from product innovations. As a product innovation, iPod's success is based on solid foundations in science, math, and engineering knowledge, however, the majority of its value-added comes from Apple's creative marketing as well as business innovations in response to global consumer needs and expectations (Linden et al. 2007). Likewise, Google's success in new product innovation is mostly sourced by its capacity in technology development using science, engineering, and mathematics knowledge, however the market success of these new products becomes possible when consumers evaluate them as having a value worth paying for. A product innovation exploit market value whenever consumers perceive that it satisfies one of their needs even the one that they have not been aware before. Therefore, in today's business environment for successful innovations what matters most have never been only proficiency in scientific know-how, but what makes the difference emanate mostly from having a comprehension of basic human needs and expectations.

Today's changing nature of innovations also leads to a wider range of skill requirements for today's workforce. Explicitly, skill set requirement for working in Industry 4.0 requires more than core hard-skills, though it constitutes quite a wide mix of skills including soft-skills as communication, coordination and autonomy skills. Therefore, it is not just developing 'a digital workforce' for future, but it is being able to develop a future workforce who is capable of seeing the 'big picture' by realizing their organization as a whole, together with identifying dynamics of their industry, besides how different industries and organizations interrelate in terms of their value chains. According to the projections of World Economic Forum, the shifts in skill requirements of employees are likely to continue in future (2016a). In Future of Jobs Report (2016a), it is pointed that at 2015 the top ten skills required from employees in order of importance are as follows: (1) complex problem solving, (2) coordinating with others, (3) people management, (4) critical thinking, (5) negotiation, (6) quality control, (7) service orientation, (8) judgment and decision making, (9) active listening, and (10) creativity. However, it is projected that as of 2020, there will be a shift in employee skill requirements, and the top ten skills according to their order of importance would be as follows: (1) complex problem solving, (2) critical thinking, (3) creativity, (4) people management, (5) coordinating with others, (6) emotional intelligence, (7) judgment and decision making, (8) service orientation, (9) negotiation, and (10) cognitive flexibility. According to the predictions of World Economic Forum based on data obtained from chief human resources and strategy officers from leading global employers about what the current shifts mean specifically for employment, skills and recruitment across industries and geographies, it has seen that soft-skills including people management, coordinating with others, emotional intelligence, and negotiation would become very important both for success of companies and individual employees.

7.3   Talent Development Practices for Industry 4.0

Product innovations are the driver force for today's economy, however, in order to make money from these product innovations, organizations need to attain market exploitation which only happens when consumers associate value to these products that is worth buying. Therefore, in today's competitive and resourceful work context, organizations need employees who not only have proficiency in scientific know-how to be able to make product innovations, but also have comprehension of human nature that would help them to recognize consumer needs so that these innovations would have markets to be sold. That is to say, in order to utilize the benefits of Industry 4.0 and adapt to the changing nature of innovations, organizations need employees who can understand, use, and develop novel work designs. Recent digital transformation brings about a range of new roles within organizations, which necessitate a new set of employee skills, as explained in the earlier section. Therefore, it is not just developing 'a digital workforce' for future, but it is being able to develop a future workforce that is capable of seeing the 'big picture', adapting immediately, and thinking and working innovatively, so that its contribution would create a competitive edge for organization.

In line with progressively increased augmentation of work systems, necessity for creating future workforce with necessary skills is becoming ever more urgent. By being the first, but not the last, manufacturing companies are having difficulty filling jobs with employees having required skills. Although there are relatively high unemployment rates all over the world, the current skill sets of employees may not be adequate enough for the emerging work designs and job roles. The problem is realized mostly as finding the right talent for the right job. As a result, there is an increasing competition for talent within a range of industries, primarily in manufacturing as being a sector which has been traditionally viewed as not offering the most remunerative and attractive professional career prospects for young generations all along.

According to a PwC report (2015) based on 8th Annual Global CEO survey, 73% of CEOs cite skill shortages as a threat to their businesses, and 81% say they are looking for a much broader range of skills when hiring than in the past. In line with the widening digital skill gap, competition over talented employees, not only for today's economy but also for tomorrow's further digitalized businesses, has presently intensified. Facing the challenge of creating an adaptable workforce within the labor market not having adequate amount of digital talent, a growing number of companies have been working on formulating their talent development strategies together with their digital strategies. Obviously seen by all, what companies really need is not only a business strategy, but it's more about having the right talent who would successfully implement that strategy.

Talent and talent management concepts have been quite popular since the early 90s, together with the famous article 'The war for talent' by McKinsey & Co consultants (Chambers et al. 1998). Their study highlighted the important role of employees in achieving outstanding company performance, also validated employees' significant impact on companies' competitive advantages. Thereafter, the concept of 'the war for talent' referred for the situation where employers compete with each other to recruit and retain valuable employees. On the subject of strategic management, the importance of talent management practices derive from the fact that talented employees have the strategic capabilities that can increase the productivity, efficiency and competitive advantage of organizations in various industries (Khilji et al. 2015). This is in line with the assertions of resource-based view of the firm (e.g., Conner and Prahalad 1996), which claims that "the value of human capital is inherently dependent upon its potential to contribute to the competitive advantage or core competence of the firm" (Lepak and Snell 1999, p. 35). For that reason, talent development of a company should be positioned as a strategic level decision, and talent management practices need to be aligned with company's overall strategy together with all strategic business processes.

In broad terms, 'talent' mainly refers for individuals having skills, intelligence, and capability in some profession that permit performing specific acts in a superior level. However, as a concept, talent incorporates a variety of meanings, and thus, it seems quite reasonable to have a lack of consistent definition for talent in the current talent management literature. Among various definitions of talent in the literature, the one by Ulrich and Smallwood (2012) stands out by its broader scope, that is: *"Talent = competence [knowledge, skills and values required for today's' and tomorrows' job; right skills, right place, right job, right time] × commitment [willing to do the job] × contribution [finding meaning and purpose in their job]"* (p. 60). On the other hand, talent management process involves acts of identification, attracting, developing, rewarding and retaining employees with critical attributes by which these employees would contribute to the sustainability of organizational success as well as organizational development (Collings and Mellahi 2009). Talent management practices enable companies to utilize, develop, and manage talented employees.

As in the past, also today companies need talented employees to achieve superior organizational performance and competitiveness; however, the identification of strategic competencies that characterize 'who the talented employees are' have been changing together with the novel work models derived from digitalization by Industry 4.0. Moreover, globalization together with the digital transformations lead to progressive disappearance of boundaries between countries, economies, and so organizations, through which employees have started developing 'boundaryless careers', which implies for less structured employment opportunities beyond borders and structures (Baruch 2003). Resultant employee mobility that becomes possible not only across organizations but also across jobs, industries, geographies, and employment contracts brings different employment opportunities for employees, while it also challenges organizations by broadening the scope of their competition for talent. More flexible organizational structures together with increased employee mobility lead to the gradual disappearance of well-defined career ladders. From employers' point of view, given that they are no longer willing or able to provide continued employment assurance to their employees due to rapid

changes for required employee skills and transformations in organizational structures to a more flexible and flatten formations, they have supported employees'
initiative for taking control of their own careers. In view of that, the focus of talent
development practices has gradually shifted from being solely internal to the one
that integrates external and internal sources (e.g., Piore 2002). As the competition
for talent intensifies due to the widening global talent shortage by the increased skill
mismatch between skills supply and demand in the labor market (World Economic
Forum 2016a), the ability to attract, motivate and retain talent, becomes gradually
more crucial for the sustainability of organizational success. In this new competitive
talent landscape, there is uncertainty both in the demand for and supply of talent
(Cappelli 2008).

In Industry 4.0 era, companies need to start their talent development processes
with a basic question of *'which skills will be highly needed now and in the future
specifically for their business, in their industry and all over the global economy?'*.
To be able to answer this question, companies should begin with making data
collection and comprehensive data analyses on their current and future work
designs, so that they could understand which roles will be automated in their
industry, while which new roles will be required in line with ongoing automations?
After that, for each of the work roles, the necessary skill sets need to identified
through comparative studies that involve past practices and benchmark applications
within and across sectors. As a result, companies identify the strategic competencies
that characterize *'who the talented employees are'* for their industry and business,
and so they would know what skills should they be looking for as well as training
their current employees for. The talent development planning should be made in
guidance by comprehensive data analysis of job designs and performance outcomes
so that they would verify which jobs make a difference for the success of organization. In other words, for the *'strategic jobs'*, i.e., jobs which are more critical to
organizational performance than others, companies should devote more resources to
them in which individual performance has the greatest potential to impact firm
performance (Huselid et al. 2005). Integration of the focus of talent development
from jobs and skills perspective depends on the fact that value of a superior individual performance is often moderated by the job being occupied as stated by
Becker and Huselid (2006, p. 904), *"The value of employee skills within a firm is
not just a supply side phenomenon. It is a function of how those skills are used and
where they are used."*

As a next step in talent development processes, companies should make conversion of their talent development plans into integrated set of activities by taking
necessary steps for combining the right people having the necessary skills with the
jobs which are highly critical. In that, for today's talent development practices
deciding on how to attract and retain employees with the required skill sets requires
a special focus. Together with increased mobility of employees via globalization
and digitalization, attracting and retaining high talent becomes highly challenging
for organizations. As a result of changing demographics in labor markets all over
the world which is mostly shaped by the fresh entrances to talent pools from

generations of Millennials and Gen Z, a new set of principles is emerging for attracting and retaining talent. In less than a decade, Millennials together with Gen Z would dominate the future workforce, and unlike the generations preceding them they prefer having greater flexibility and autonomy rather than fixed employment contracts. According to a World Economic Forum survey, career advancement (48%), company culture (38%) and training/development opportunities (32%) are what millennials are looking to get from their employer (Cann 2015). Therefore, for attracting and retaining these young talent, companies should focus on positioning themselves as an employer of choice via being and branding as great places to work. This requires adopting more flexible employment contracts since permanent employment status is no longer the sole preference, but temporary-base or project-based contracts are also increasing as alternative contracting methods (ILO 2015).

In addition to offering flexible employment alternatives, companies can attract and retain digital talent by improving their company culture and offering incentives that are relevant in the digital age. Creating a workplace that attracts talented employees requires a progressive and forward-looking organizational culture driven by a strong and visionary leadership. Today's effective leadership definition is moving away from autocratic top-down approaches towards more authentic and collaborative styles. Leaders need to be capable of instilling a corporate culture where digital systems can further flourish, accompanied by a company vision derived from solid knowledge about technology and how it is disrupting their businesses.

Such kind of open and visionary leadership approaches resonate well with millennials since they have grown up in an environment where data and people are highly accessible. This new generation is used to be part of a highly integrated social media and to share and receive instant data about themselves, individuals, companies and products. They align themselves with technology since they all born as digital natives. Accordingly, companies may consider more utilizing social media and online platforms for attracting these new talent. In the same way, organizations should try to integrate the most updated technologies into their daily ways of doing work by the tools and hardware preferences. Moreover, companies should create suitable work environments together with novel work designs where humans and robots can successfully work together. If necessary, work processes should be re-designed for reducing the skill mismatch between jobs and employees.

In order to ensure the sustainability of these talent development processes, companies need to pay attention constantly developing their workforce including the newly acquired talent together with their existing employees. Offering continuous training and learning opportunities as well as aligning organization's key performance indicators with employee incentive plans maintain the utility of talent development practices. In line with the preferences of new generations, it is critical to take a practical as well as long-term view in their training practices by offering contemporary training options such as massive online courses (MOOCs).

7.4   Conclusion

In a global world, the most indispensable success factor for economic sustainability and competitive advantage of organizations is their capacity to adapt to change. Automation of business processes together with the emergence of novel business models caused by recent innovations in digital technologies have been intensely changing dynamics of most industries. In line with these transformations, the 'ways of doing' work are also changing, and these novel work systems and designs impose new skill requirements for employees.

Together with an already tight market for talent, developing future workforce and being ready for the upcoming changes in employment landscape is increasingly present in strategic planning priorities of companies. For strategic workforce planning, companies need to begin with understanding the changing scope and content of work requirements and workforce skill necessities. Subsequently, they need to evaluate the availability of talented employees with the right skills and capabilities required for future in their current organizational content. At the same time, they need to know how to attract and recruit new talent with necessary skills.

Industry 4.0 era necessitates all employees, even workers with low-skilled jobs, to have a collection of ICT skills. However, Industry 4.0 requires essential employee skill sets to entail more than core skills; indeed, for successful execution of hard-skills employees should have soft-skills as collaboration, communication and autonomy for being able to execute their jobs in hybrid operating systems. In today's complex world of work, adaptability becomes the most important capability for employees, that's why, for being successful in their jobs employees need to get into the habit of continuous learning, not simply in their own profession but in a wider angle through an interdisciplinary perspective. Moreover, innovative capacity is vital for competitive advantage of organizations, and in order to facilitate successful innovations even though proficiency in scientific know-how is a base requirement, yet what makes the difference emanate mostly from having a comprehension of consumer expectations which is possible only by understanding basic human needs. In view of that, it is not just developing 'a digital workforce' for future, but it is being able to develop a future workforce capable of seeing the 'big picture' by recognizing interrelations between different stakeholders including customers, organizations and industries as well as different value chains evolving within Industry 4.0.

As a result, developing future workforce for Industry 4.0 requires not only attracting and recruiting new talent which is needed, but also re-skilling current employees through training programs, and, if necessary, re-designing work processes to eliminate the skill mismatch between jobs and employees. Consequently, organizations should work on offering new learning experiences, creating new development opportunities, and building robust engagement systems to be used for their talent development practices.

References

Autor DH (2015) Why are there still so many jobs? The history and future of workplace automation. J Econ Perspect 29(3):3–30

Autor DH, Handel MJ (2013) Putting tasks to the test: human capital, job tasks, and wages. J Labor Econ 31(2):59–96

Autor DH, Levy F, Murnane RJ (2003) The skill content of recent technological change: an empirical explanation. Quart J Econ 118(4):1279–1333

Baruch, Y (2003) Transforming careers: from linear to multidirectional career paths organizational and individual perspectives. Career Dev Int 9(1):58–73

Becker BE, Huselid MA (2006) Strategic human resources management: where do we go from here? J Manag 32(6):898–925

Beechler S, Woodward I (2009) The global "war for talent". J Int Manag 15:273–285

Berger T, Frey CB (2016) Did the computer revolution shift the fortunes of U.S. cities? Technology shocks and the geography of new jobs, Reg Sci Urban Econ 57:38–45

Brynjolfsson E, McAfee A (2011) Race against the machine: how the digital revolution is accelerating innovation, driving productivity, and irreversibly transforming employment and the economy. Digital Frontier Press, Lexington, Mass

Cann O (2015) 3 things millennials want from work. World Econ Forum Blog. https://agenda. weforum.org/2015/10/3-things-millennials-want-from-work. Accessed 25 Oct 2015

Cappelli P (2008) Talent on demand: managing talent in an age of uncertainty. Harvard Business School Publishing, Cambridge, MA

Carnevale A, Smith N, Melton, M (2011) STEM Science, engineering, technology and mathematics. Georgetown University Center on Education and the Workforce, Washington, D.C. http://cew.georgetown.edu/STEM. Accessed Oct 2011

Chambers EG, Foulon M, Handfield-Jones M, Hankin SM, Michaels EG (1998) The war for talent. The McKinsey Quarterly

Cisco (2011) The internet of things: how the next evolution of the internet is changing everything. White Paper, Cisco Internet Business Solutions Group (IBSG), April 2011

Collings D, Mellahi K (2009) Strategic talent management: a review and research agenda. Hum Resource Manag Rev:304–313

Conner KR, Prahalad CK (1996) A resource-based theory of the firm: knowledge vs. opportunism, organization science, 7:477–501. http://dx.doi.org/10.1287/orsc.7.5.477

Frey CB, Osborne MA (2013) The future of employment: how susceptible are jobs to computerisation? Oxford Martin School

Grundke R, Squicciarini M, Jamet S, Kalamova M (2017) Having the right mix: the role of skill bundles for comparative advantage and industry performance in GVCs. OECD Science, Technology and Industry Working Papers, OECD Publishing, Paris

Guthridge M, Komm A, Lawson E (2008) Making talent a strategic priority. McKinsey Quarterly, 4

Hill, C (2007) The post-scientific society, issues in science and technology fall. http://www.issues. org/24.1/c_hill.html

Huselid MA, Beatty RW, Becker BE (2005) "A players" or "A positions"? The strategic logic of workforce management. Harvard Bus Rev 83(12):110–117

ILO (2015) World employment social outlook: the changing nature of jobs

Khilji S, Tarique I, Schuler R (2015) Incorporating the macro view in global talent management. Hum Resource Manag Rev:236–248

Lepak DP, Snell SA (1999) The human resource architecture: toward a theory of human capital allocation and development. Acad Manag Rev 24(1):31–48

Linden G, Kraemer KL, Dedrick J (2007) Who captures value in a global innovation system? The case of Apple's iPod. Irvine, CA: Personal Computing Industry Center. http://pcic.merage.uci. edu/papers/2009/InnovationAndJobCreation.pdf

McKinsey Global Institute (2017) A future that works: Automation, employment, and productivity. McKinsey & Company, January 2017

OECD (2016a) Forum 2016 Issues: the Future of Education, Available: http://www.oecd.org/forum/issues/forum-2016-issues-the-future-of-education.htm

OECD (2016b) Getting skills right: assessing and anticipating changing skill needs. OECD Publishing, Paris

Piore MJ (2002) Thirty years later: internal labor markets, flexibility and the new economy. J Manag Gov 6(4):271–279

PwC (2015) People strategy for the digital age—A new take on talent, PwC 18th Annual Global CEO Survey, 2015

Quintini, G (2014) Skills at work: how skills and their use matter in the labour market, OECD social, employment and Migration working papers, No. 158, OECD Publishing, Paris

Ulrich D, Smallwood N (2012) What is talent? Leader to Leader 63:55–61

World Economic Forum (2016a) The future of jobs report. Geneva, Switzerland, Cologny

World Economic Forum (2016b) Digital transformation of industries. Geneva, Switzerland, Cologny

Chapter 8
The Changing Role of Engineering Education in Industry 4.0 Era

Sezi Cevik Onar, Alp Ustundag, Çigdem Kadaifci and Basar Oztaysi

Abstract The new industry 4.0 era necessities new cross-functional roles with different knowledge and skills that combine IT and production knowledge. The universities and their engineering departments have a vital role in fulfilling this need. There are a number of departments offering these new engineering education requirements, but the characteristics of these departments and how they converge to and diverge from each other are yet to be revealed through objective evaluation. Such evaluation should be based on a precise classification of knowledge and skills areas offered in these departments. Therefore, it is important to understand the characteristics of knowledge and skills provided in these departments to determine the emerging patterns in the delivery of new education requirements of Industry 4.0. The main objectives of this chapter is to define the new education requirements incorporated into Industry 4.0, and reveal the emerging patterns and similarities in engineering education to cover this need. In order to address these issues we study a sample of 124 engineering departments.

8.1 Introduction

The fourth industrial revolution will make business processes more efficient and productive. On the product side, it will also extract greater value from data for usage-based design and mass customization, which in turn will open the way to new markets (Rossi 2016). However, adoption of Industry 4.0 will result in the elimination of lower skilled jobs through automation and digitalization (Unnikrishnan and Aulbur 2016). This transformation will require a significant change in workforce skills, organizational structures, leadership mechanisms and corporate culture.

S. Cevik Onar (✉) · A. Ustundag · Ç. Kadaifci · B. Oztaysi
Department of Industrial Engineering, Faculty of Management,
Istanbul Technical University, 34367 Macka, Istanbul, Turkey
e-mail: cevikse@itu.edu.tr

© Springer International Publishing Switzerland 2018   137
A. Ustundag and E. Cevikcan, *Industry 4.0: Managing The Digital Transformation*,
Springer Series in Advanced Manufacturing, https://doi.org/10.1007/978-3-319-57870-5_8

According to Ahrens and Spöttl (2015), five parameters are important for the required qualification of skilled workers in industry 4.0:

- comprehensive integration and information transparency
- increasing automation of production systems
- self-management and decision-making by objects
- digital communication and interactive management functions
- flexibilization of the use of staff.

The new era will create many new cross-functional roles for which workers will need both IT and production knowledge. Therefore, education systems should transform by providing broader skill sets and job-specific capabilities, close the IT skills gap, and offer new formats for continuing education (BCG 2015). Not only the educational contents but also the methods of skill development had to meet the requirements of a new generation of employees. Peter Fisk (2017) defines a new vision for the future of education as "Education 4.0" in a wider scope:

- which responds to the needs of industry 4.0, where man and machine align to enable new possibilities
- harnesses the potential of digital technologies, personalized data, open sourced content, and the new humanity of this globally-connected, technology-fueled world
- establishes a blueprint for the future of learning—lifelong learning—from childhood schooling to continuous learning in the workplace, to learning to play a better role in society.

Peter Fisk (2017) defines the new characteristics of future education as personalized, repackaging, peer-to-peer and continuous based on the Clay Christensen's ideas. According to Fisk, the learning process will change the old mindsets in the near future whether it is classroom or workplace, online or offline, structured or unstructured, taught or learned, standardized or not, certificated or not.

Universities will also have a significant role in the workforce transformation in the new era. To encourage cross-functional knowledge and communication, they should change their existing engineering programs. According to Jeschke Sabina (2015), engineering education requires the collaboration of academic disciplines with the development of highly complex, socio-technical systems. Future engineers need the skills to adapt the innovation cycles rapidly, and IT is the primary driver of innovation in future industrial context. In this regard, universities should focus on building specific capabilities for the new roles and adapting their curricula to meet companies' expectations for Industry 4.0 skills. They also need to encourage soft skills that enable workers to be open to ongoing capability development, inter-disciplinary collaboration, and innovation (BCG 2015).

Since the big data, cloud computing, data analytics, artificial intelligence, machine learning, IoT systems, adaptive robotics, virtualization and additive manufacturing are the main technology components of Industry 4.0, all these subjects should be included and taught comprehensively in the engineering education programs.

The university programs should be restructured so that the innovation and entrepreneurship management skills of the engineers will be developed. Additionally, e-learning technologies such as gamification, virtual labs, and learning analytics should support the engineering education process.

The main question for the universities is how the structures and course contents of different engineering programs such as computer, electronics, industrial and manufacturing will be changed in the new era. The main objective of this study is to investigate the new requirements for the engineering education programs of universities in industry 4.0 era.

The following section of this chapter explore the new education requirements in industry 4.0 era. Section 8.3 reveals the relation among current engineering education and the new engineering education requirements. The last section concludes and gives further suggestions.

8.2 New Education Requirements

In this subsection, new education requirements for Industry 4.0 are summarized in three groups. First, the content requirements for education in Industry 4.0 era is given, then in the second stage requirements and advances in educational technologies are summarized and finally the role of ability to work interdisciplinary teams is emphasized.

8.2.1 Education Content

The content requirements for education in Industry 4.0 era can be classified into four main groups. The first group is composed of data collection, storage, and processing technologies. The second group focuses on value added automated operations. The third group is composed domain knowledge, which includes the state-of-the-art applications in specific domains and the final group is innovation and entrepreneurship.

8.2.1.1 Data and Computing Technologies

In recent years, various new technologies have emerged for collecting, storing and processing data to produce useful applications. Some of the most important technological advances related to Industry 4.0 education can be listed as follows:

- Data modeling and Big data: Data is vital for many IT applications, so data modeling which deals with efficient data storage for traditional data sources is an important topic. Recently, big data which, refers to the large volume of structured and unstructured data has become a crucial area with its specific technologies.

- Data Analytics: Data Analytics is the process of extracting meaning from raw data using specialized computer systems. There are three main groups of data analytics, descriptive analytics deal with data processing for maintaining a summary of historical data to produce useful information. Predictive analytics focus on identifying the likelihood of future outcomes of potential actions using specifics algorithms and historical data. Finally, prescriptive analytics tries to find the best path forward based on the constraints and objectives of a problem.
- Cloud computing: Cloud computing is defined as on-demand delivery of computational power, data storage resources, software, and other IT resources through a platform via the internet. Cloud computing enables centralized information for industry 4.0 applications and offers a platform collaboration to advance and refine research for entire industry gains.
- Machine Learning: Machine Learning is the subfield of computer science that gives "computers the ability to learn without being explicitly programmed" (Samuel 1959). Machine learning is important for industry 4.0 since it enables autonomy to software.

8.2.1.2 Value Added Automated Operation

Value added automated operations involve research areas such as robotic & automation, smart and embedded systems, and additive manufacturing.

- Automation: Automation is the use of several control systems for operating equipment. Industrial automation in manufacturing refers to utilizing intelligent machines in factories in order to minimize human intervention in manufacturing processes.
- Robotics: Robotics is a branch of engineering that involves the conception, design, manufacture, and operation of robots. Robotics is necessary for Industry 4.0 since using industrial robots for production provide high efficiency and flexibility.
- Smart and embedded systems: Internet of things (IOT) is the internetworking of physical entities that enable them to collect and exchange data using electronics, software, sensors, actuators, and network connectivity.
- Additive manufacturing: Additive manufacturing refers to processes used to create a three-dimensional object in which layers of material are formed. It is a key component in Industry 4.0 since it enables the production of the desired components faster, more flexibly and more precisely.

Besides, the above-mentioned technological contents students should be equipped with specific and up-to-date domain knowledge. The students should know the state-of-the-art industry 4.0 applications and current business requirements in industries such as energy, healthcare, and service.

8.2.1.3 Innovation and Entrepreneurship

Innovation and entrepreneurship have been vital for economic development. In the era of Industry 4.0, this role gets even more critical, and education needs to be aligned with this change. As Jeschke (2015) implies, companies need to implement innovation process management to cope with competition and survive the financial crisis. On the other hand, as the innovation cycles become faster, a need for more enterprises emerges.

Innovations are divided into two categories; evolutionary innovations are continuous and incremental advances in technology or processes. Revolutionary innovations on the other hand, are entirely new and disruptive changes in technology and processes. In times of industrial revolutions, revolutionary innovations dominate which means disruptive technologies and processes are invented. In the scope of this section, the question becomes how to teach the students so that they become innovators. Innovation and entrepreneurship can be classified under two research areas as follows:

- Innovative materials and tools: Innovation and entrepreneurship can be related to the new materials/tools that enable developing new products. These new materials such as Nano-technological materials are at the heart of the Industry 4.0.
- Innovation/Entrepreneurship business models: Innovation management has been changed. Lately, open innovation and crowdsourcing have become very popular. Open innovation suggests that companies should use external ideas when they look for technological advances. New education for Industry 4.0 should focus on these new concepts. Besides the classical entrepreneurship skills, the students need to be equipped with novel skills such as new business thinking, collaboration with different parties, and communication with various stakeholders. Risk taking and decision making under uncertainty, creativity, and ability to adapt to rapid changes are also essential skills of entrepreneurs in the age of Industry 4.0.

8.2.2 E-Learning Technologies

Technologies regarding Industry 4.0 education can be classified into three main groups. The first group is related with virtual labs and augmented reality for education. The second group is using gamification for education and the third group focus on learning analytics.

- Virtual Labs and Augmented Reality: Virtual labs refer to software for interactive learning based on simulation of real phenomena. It allows students to explore a topic by comparing and contrasting different scenarios, to pause and restart the application for reflection and note-taking, to get practical experimentation experience over the Internet (Igi-Global 2017). Augmented reality, on the other hand, is a live direct or indirect view of a physical, real-world environment whose elements are augmented by computer-generated sensory input such as sound, video, graphics or GPS data. Both of the technologies can be used for educational purposes empowering interactivity between the system and the learner. Particularly in cases where building real Industry 4.0 laboratories are expensive or impossible, the interactivity maintained by these systems may foster effective learning.

- Gamification: Gamification is the application of game-design elements and game principles in non-game contexts. Typically, gamification tries to improve user engagement and organizational productivity. Gamification is also important for learning systems since it captures the attention of the learner via game principles such as; narrative, instant feedback, leveling up, and progress indicators. With gamification, real life challenges can be mimicked with increasing levels of difficulty and social learning can be maintained by promoting social interaction and competition.
- Learning Analytics: Learning analytics (LA) is the application of data analytics into e-learning environments. LA measures and collects data about the learners and their context in order to understand the level of learning and optimize the future actions. With learning analytics applications, the learning process can be personalized, adaptive content can be produced, learner achievement can be enhanced, and teachers become more effective. As Industry 4.0 contain various different research areas and application fields enabling adaptive and efficient learning is very important. With a personalized learning system, students may get detailed knowledge on specific area they want to focus on.

8.2.3 Working in Interdisciplinary Teams

Although Industry 4.0 is associated with state-of-the-art technologies, human perspective is also critical and bring important challenges. Industry 4.0 involves many research area including, mechatronic engineering, industrial engineering, and computer science. This nature of Industry 4.0 brings the need for working in interdisciplinary teams, realizing interdisciplinary tasks and provide interdisciplinary thinking. In traditional universities, which focus on a single engineering discipline, it is not easy to equip the students with interdisciplinary skills required for Industry 4.0. Program structures and/or course syllabuses may be updated to improve the interdisciplinary skills.

8.3 New Engineering Education Requirements and the Current Engineering Education

Building on the new education requirements, we review the research areas of a sample of covers industrial engineering, mechanical engineering, electrical and electronics engineering, and computer engineering disciplines. We analyze the research areas of these disciplines with respect to the main and sub new education defined in Sect. 8.2. We follow a two-stage approach. First, we identify the sample programs and collect the research areas of these programs. We want to reveal the research area differences in different cultures. Therefore, a diversified sample is selected. Although some research areas incorporate with more than one educational requirement, by considering the content we assign the research area into one of the groups. We analyze a total of 124 programs which covers industrial engineering, mechanical engineering, electrical and electronics engineering, and computer engineering disciplines delivered at 32 universities (see Table 8.1) in eight countries, namely the USA (19 universities), Turkey (five universities), China

Table 8.1 List of universities

Abb.	University	Abb.	University
Berkeley	University of California, Berkeley	Princeton	Princeton University
Bilkent	Bilkent University	Purdue	Purdue University
BOUN	Bogazici University	RWTH	RWTH Aachen University
Columbia	Columbia University	Sabancı	Sabancı University
Cornell	Cornell University	Seoul	Seoul National University
Eindhoven	Eindhoven University of Technology	Stanford	Standford University
Georgia	Georgia Institute of Technology	Texas A&M	Texas A&M University
Hong Kong	Hong Kong University of Science and Technology	Texas Austin	The University of Texas at Austin
Koç	Koç University of Science and Technology	Tokyo	Tokyo Institute of Technology
Lehigh	Koç University	Tsinghua	Tsinghua University
MIT	Massachusetts Institute of Technology	UFL	University of Florida
Northwestern	Northwestern University	Illinois	University of Illinois Urbana-Champaign
NUS	National University of Singapore	UMICH	University of Michigan–Ann Arbor
METU	Middle East Technical University	USC	University of Southern California
Ohio State	The Ohio State University	Wisconsin	University of Wisconsin-Madison
Penn State	Penn State University	VT	Virginia Tech

(two universities), Germany, Singapore, South Korea, Japan, and The Netherlands. We classify the programs based on their closeness to the discipline. In the universities, the name of the departments show a high variety, we classify the programs based on their closeness to the disciplines. For example, Princeton University's Operations research & financial engineering department is classified under industrial engineering discipline. The full list of the departments is given in Appendix A.

In the second stage, we classify the research areas according to the new education requirements defined in Sect. 8.2. Once we determine which education requirement the program is offering, we are able to ascertain the configuration of the programs and differences among disciplines. We provide the classification of universities research areas into the three new education requirements that we have determined from the literature. We also derive sub-education requirements from our analysis.

8.3.1 Innovation/Entrepreneurship

The first new education requirement is entrepreneurship and innovation. This education requirement consists mainly two distinct requirements. The first one focuses on the usage of innovative materials in the new technology development whereas the second one focuses on the innovation and entrepreneurship with a business perspective. In Fig. 8.1, we give the departments that have special research areas considering the innovation and entrepreneurship. Different engineering disciplines are marked with different shades such as industrial engineering discipline is represented with the darkest shade whereas electrical and electronics engineering is represented with the lightest shade. Each department that has a focus on the sub-knowledge area is represented with a rectangle. The size of the rectangle is proportional to the number of research areas in the field.

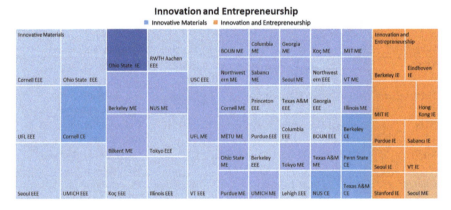

Fig. 8.1 Innovation and entrepreneurship

According to the Fig. 8.1, innovation and entrepreneurship knowledge area with a business perspective is dominated by industrial engineering discipline. Nine industrial engineering departments are focusing on this area. Innovative materials knowledge area is dominated by mechanical engineering and electrical and electronics engineering disciples. Out of 32 mechanical engineering departments in 20 departments there exist a program related to innovative materials. Only four computer engineering programs have a research area on innovative materials.

8.3.2 Data and Computing Technologies

Data and computing technologies involve data analytics, signal and image processing, data modeling and big data, machine learning, networks, cyber security and cloud computing sub-education requirements.

In Fig. 8.2, we give the departments that have special research areas considering the data and computing technologies. Similar to the Fig. 8.1, different engineering disciplines are marked with different shades. The sub-knowledge areas are colored with different colors. Each department that has a focus on the sub-knowledge area is represented by a rectangle. The size of the box is proportional to the number of research areas in the field.

Figure 8.2 shows that almost every discipline has an interest in data and computing technologies. Networks and network design sub-knowledge area is dominated by electrical and electronics engineering and computer engineering disciplines. Yet, several industrial engineering programs are interested in this field. Similarly, the machine learning sub-learning area is dominated by electrical and electronics engineering and computer engineering disciplines. Six mechanical engineering and two industrial engineering departments are focusing on machine learning knowledge area. Data modeling and big data knowledge area is a

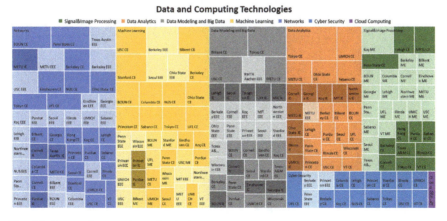

Fig. 8.2 Data and computing technologies

diversified field where computer engineering, industrial engineering and electrical and electronics engineering disciplines have almost equal impact. Computer engineering and industrial engineering disciplines dominate data analytics. Computer engineering discipline has a high impact on all data computing technologies including cyber security. Signal and image processing is the only sub-knowledge area where mechanical engineering discipline has a more powerful effect.

8.3.3 Value Added Automated Operations

Value added automated operations education requirement covers gaining knowledge on the additive manufacturing, automation, robotics, and smart and embedded systems.

In Fig. 8.3, we give the departments that have specialized research areas considering the value added automated operations. In this figure, different engineering disciplines are marked with different shades and the sub-knowledge areas are colored with different colors. Each department that has a focus on the sub-knowledge area is represented by a box. The size of the box is proportional to the number of research areas in the field.

Figure 8.3 shows that every discipline accepts industrial engineering has an interest in value-added automated operations. Robotics field is covered by all these three disciplines. Computer engineering and electrical and electronics engineering disciplines dominate smart and embedded products knowledge areas. Automation field is mainly dominated by mechanical engineering. Only one computer engineering, one electrical, and electronics engineering department focus on this area.

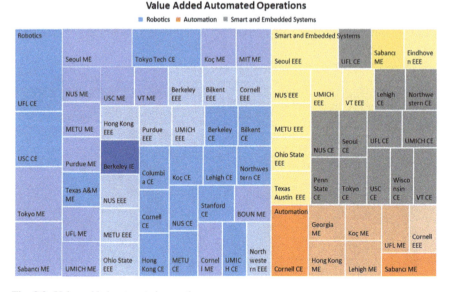

Fig. 8.3 Value added automated operations

When we consider all the engineering departments, they show a wide variety of research areas. Despite some common characteristics, the nature of engineering departments in these new education requirements are changing on the faculty members. Cultural differences among the countries do not exist in engineering schools.

8.4 Conclusion and Further Suggestions

The shortage of new engineering education requirements is covered by a number of engineering education alternatives such as industrial, computers, electric & electronics and mechanical engineering disciplines. Although there are some common characteristics, engineering research areas appear to be designed on a rather ad hoc basis, depending on the availability of faculty and their research areas. This has led to an unbalanced representation of new education requirements since the available resources tend to be limited.

The new Industry 4.0 era teaching spans multiple disciplines and uses an increasing amount of methods that demonstrate the complexity associated with this growing discipline. There is a strong connection between the new education requirements of industry 4.0. Therefore, in order to achieve practical information and knowledge applicable to the business environment, different disciplines should be able to work collectively. Designing integrated engineering programs that cover these new engineering education requirements may close the gap between the universities and the business environment.

The contribution of this research lies within its objective approach to departments and the comprehensive, revealing process used in the recent status of the new engineering requirements. Our chapter analyses engineering departments from the knowledge and skill areas perspective and clusters them with respect to their offered research area. This research can be a guide in the design process of new engineering programs that satisfies the needs of Industry 4.0 era.

Appendix A

Abb.	Department
Berkeley IE	University of California, Berkeley—Industrial Engineering and Operations Research
Bilkent IE	Bilkent University—Industrial Engineering
BOUN IE	Bogazici University—Industrial Engineering
Columbia IE	Columbia University—Department of Industrial Engineering and Operations Research
Cornell IE	Cornell University—Operations Research and Engineering

(continued)

(continued)

Abb.	Department
Eindhoven IE	Eindhoven University of Technology—Industrial Engineering & Innovation Sciences
Georgia IE	Georgia Institute of Technology—Industrial Engineering
Hong Kong IE	Hong Kong University of Science and Technology—Industrial Engineering and Logistics Management
Koç IE	Koç University of Science and Technology—Industrial Engineering
Lehigh IE	Koç University—Industrial and Systems Engineering
MIT IE	Massachusetts Institute of Technology—Sloan School of Management
Northwestern IE	Northwestern University—Industrial Engineering & Management Sciences
NUS IE	National University of Singapore—Industrial and Systems Engineering
METU IE	Middle East Technical University—Industrial Engineering
Ohio State IE	The Ohio State University—Department of Integrated Systems Engineering
Penn State IE	Penn State University—Department of Industrial Engineering
Princeton IE	Princeton University—Operations Research & Financial Engineering
Purdue IE	Purdue University—School of Industrial Engineering
RWTH IE	RWTH Aachen University—Industrial Engineering
Sabancı IE	Sabancı University—Industrial Engineering
Seoul IE	Seoul National University—Industrial Engineering
Stanford IE	Standford University—Management Science and Engineering Program
Texas A&M IE	Texas A&M University—Department of Industrial and Systems Engineering
Texas Austin IE	The University of Texas at Austin—Operations Research & Industrial Engineering
Tokyo IE	Tokyo Institute of Technology—Industrial Engineering and Economics
Tsinghua IE	Tsinghua University—Industrial Engineering
UFL IE	University of Florida—Department of Industrial and Systems Engineering
Illinois IE	University of Illinois Urbana-Champaign—Department of Industrial and Enterprise Systems Engineering
UMICH IE	University of Michigan–Ann Arbor—Industrial and Operations Engineering
USC IE	University of Southern California—Industrial and Systems Engineering
Wisconsin IE	University of Wisconsin-Madison—Industrial & Systems Engineering
VT IE	Virginia Tech—Department of Industrial and Systems Engineering
Berkeley ME	University of California, Berkeley—Mechanical Engineering
Bilkent ME	Bilkent University—Mechanical Engineering
BOUN ME	Bogazici University—Mechanical Engineering
Columbia ME	Columbia University—Mechanical Engineering
Cornell ME	Cornell University—Mechanical and Aerospace Engineering
Eindhoven ME	Eindhoven University of Technology—Mechanical Engineering
Georgia ME	Georgia Institute of Technology—Mechanical Engineering

(continued)

(continued)

Abb.	Department
Hong Kong ME	Hong Kong University of Science and Technology—Mechanical Engineering
Koç ME	Koç University of Science and Technology—Mechanical Engineering
Lehigh ME	Koç University—Mechanical Engineering
MIT ME	Massachusetts Institute of Technology—Mechanical Engineering
Northwestern ME	Northwestern University—Mechanical Engineering
NUS ME	National University of Singapore—Mechanical Engineering
METU ME	Middle East Technical University—Mechanical Engineering
Ohio State ME	The Ohio State University—Mechanical Engineering
Penn State ME	Penn State University—Mechanical Engineering
Princeton ME	Princeton University—Mechanical Engineering
Purdue ME	Purdue University—Mechanical Engineering
RWTH ME	RWTH Aachen University—Mechanical Engineering
Sabancı ME	Sabancı University—Mechatronics Engineering
Seoul ME	Seoul National University—Mechanical Engineering
Stanford ME	Standford University—Mechanical Engineering
Texas A&M ME	Texas A&M University—Mechanical Engineering
Texas Austin ME	The University of Texas at Austin—Mechanical Engineering
Tokyo ME	Tokyo Institute of Technology—Mechanical Engineering
Tsinghua ME	Tsinghua University—Mechanical Engineering
UFL ME	University of Florida—Mechanical Engineering
Illinois ME	University of Illinois Urbana-Champaign—Mechanical Engineering
UMICH ME	University of Michigan–Ann Arbor—Mechanical Engineering
USC ME	University of Southern California—Mechanical Engineering
Wisconsin ME	University of Wisconsin-Madison—Mechanical Engineering
VT ME	Virginia Tech—Mechanical Engineering
Berkeley EEE	University of California, Berkeley—Electrical Engineering and Computer Sciences
Bilkent EEE	Bilkent University—Electrical and Electronics Engineering
BOUN EEE	Bogazici University—Electrical and Electronics Engineering
Columbia EEE	Columbia University—Electrical Engineering
Cornell EEE	Cornell University—Electrical and Computer Engineering
Eindhoven EEE	Eindhoven University of Technology—Electrical Engineering
Georgia EEE	Georgia Institute of Technology—Electrical Engineering

(continued)

(continued)

Abb.	Department
Hong Kong EEE	Hong Kong University of Science and Technology—Electronic Engineering
Koç EEE	Koç University of Science and Technology—Electrical and Electronics Engineering
Lehigh EEE	Koç University—Electrical Engineering
MIT EEE	Massachusetts Institute of Technology—Electrical Engineering and Computer Sciences
Northwestern EEE	Northwestern University—Electrical Engineering
NUS EEE	National University of Singapore—Electrical Engineering
METU EEE	Middle East Technical University—Electrical and Electronics Engineering
Ohio State EEE	The Ohio State University—Electrical and Computer Engineering
Penn State EEE	Penn State University—Electrical Engineering
Princeton EEE	Princeton University—Electrical Engineering
Purdue EEE	Purdue University—Electrical Engineering
RWTH EEE	RWTH Aachen University—Electrical Engineering, Information Technology and Computer Engineering
Sabancı EEE	Sabancı University—Electronics Engineering
Seoul EEE	Seoul National University—Electrical and Computer Engineering
Stanford EEE	Standford University—Electrical Engineering
Texas A&M EEE	Texas A&M University—Electrical and Computer Engineering
Texas Austin EEE	The University of Texas at Austin—Electrical and Computer Engineering
Tokyo EEE	Tokyo Institute of Technology—Electrical and Electronic Engineering
Tsinghua EEE	Tsinghua University—Electrical Engineering
UFL EEE	University of Florida—Electrical Engineering
Illinois EEE	University of Illinois Urbana-Champaign—Electrical Engineering
UMICH EEE	University of Michigan–Ann Arbor—Electrical Engineering
USC EEE	University of Southern California—Electrical Engineering
Wisconsin EEE	University of Wisconsin-Madison—Electrical and Computer Engineering
VT EEE	Virginia Tech—Electrical Engineering
Berkeley CE	University of California, Berkeley—Computer Sciences
Bilkent CE	Bilkent University—Computer Engineering
BOUN CE	Bogazici University—Computer Engineering
Columbia CE	Columbia University—Computer Engineering
Cornell CE	Cornell University—Electrical and Computer Engineering
Eindhoven CE	Eindhoven University of Technology—Computer Science and Engineering
Georgia CE	Georgia Institute of Technology—Computer Engineering

(continued)

(continued)

Abb.	Department
Hong Kong CE	Hong Kong University of Science and Technology—Computer Engineering
Koç CE	Koç University of Science and Technology—Computer Engineering
Lehigh CE	Koç University—Computer Engineering
Northwestern CE	Northwestern University—Computer Engineering
NUS CE	National University of Singapore—Computer Engineering
METU CE	Middle East Technical University—Computer Engineering
Ohio State CE	The Ohio State University—Computer Science and Engineering
Penn State CE	Penn State University—Computer Science and Engineering
Princeton CE	Princeton University—Computer Science
Purdue CE	Purdue University—Computer Engineering
Sabancı CE	Sabancı University—Computer Science and Engineering
Seoul CE	Seoul National University—Computer Science and Engineering
Stanford CE	Standford University—Computer Science
Texas A&M CE	Texas A&M University—Computer Engineering
Tokyo CE	Tokyo Institute of Technology—Computer Sciences
UFL CE	University of Florida—Computer Engineering
Illinois CE	University of Illinois Urbana-Champaign—Computer Engineering
UMICH CE	University of Michigan–Ann Arbor—Computer Sciences and Engineering
USC CE	University of Southern California—Computer Engineering and Sciences
Wisconsin CE	University of Wisconsin-Madison—Computer Engineering
VT CE	Virginia Tech—Computer Engineering

References

Ahrens D, Spöttl G (2015) Industrie 4.0 und Herausforderungen für die Qualifizierung von Fachkräften. In: Hirsch-Kreinsen Hartmut, Ittermann Peter, Niehaus Jonathan (eds) Digitalisierung industrieller Arbeit. Nomos, Baden-Baden, pp 185–203

Boston Consulting Group (BCG) (2015) Man and Machine in Industry 4.0

Fisk P (2017) Education 4.0. http://www.thegeniusworks.com/2017/01/future-education-young-everyone-taught-together/

IGI-Global (2017) Online dictionary. http://www.igi-global.com/dictionary/virtual-lab/31699?, accessed at 05.04.2017

Jeshke S (2015) Retrieved from: http://www.ima-zlw-ifu.rwth-aachen.de/fileadmin/user_upload/INSTITUTSCLUSTER/Publikation_Medien/Vortraege/download//EngEducationInd4.0_22Sept2015.pdf

Rossi B (2016) Retrieved from: http://www.information-age.com/business-security-impacts-industry-4-0-123463772/

Samuel AL (1959) Some studies in machine learning using the game of checkers. IBM J 3:35–554

Unnikrishnan, Aulbur (2016) Retrieved from: http://economictimes.indiatimes.com/brics-article/skill-development-for-industry-4-0/brics_show/54460851.cms

Part II
Technologies and Applications

Chapter 9
Data Analytics in Manufacturing

M. Sami Sivri and Basar Oztaysi

Abstract Development of technology has emerged a new concept, Industry 4.0. It has come with two technological improvements, Cyber-Physical System (CPS) and Internet of Things (IoT) that drive manufacturing companies to Data Analytics by generating the huge amount data. In terms of Industry 4.0, data analytics focus on "what will happen" rather than "what has happened". These problems are entitled as predictive analytics and aims at building models for forecasting future possibilities or unknown events. The aim of this paper is to give insight about these techniques, provide applications from the literature and show a real world case study from a manufacturing company.

9.1 Introduction

Development of technology has escorted industries from mechanical systems to highly automated and smart factories in which smart machines and sensor networks. With the development of these technologies, a new concept, Industry 4.0 was introduced by German during the Hannover Fair event in 2011, which symbolizes the beginning of the 4th industrial revolution (Qin et al. 2016).

The first three industrial revolutions have resulted radical changes in manufacturing which are mechanization, using electricity and using information technology (IT). The Industry 4.0 comes with two technological improvements, Cyber-Physical Systems (CPS) and Internet of Things (IoT) that are advanced within the last decade.

Cyber-Physical Systems (CPS) is defined as transformative technologies for managing interconnected systems between its physical assets and computational capabilities (Lee et al. 2015a). The framework of this interconnectivity between sensors and networked machines is called as Internet of Things (IoT). By using IoT, various signals such as vibration, pressure, fuel consumption etc. can be extracted. Consequently, the growing use of sensors and networked machines has resulted in

M. Sami Sivri (✉) · B. Oztaysi
Istanbul Technical University, Macka, Sisli, 34367 Istanbul, Turkey
e-mail: msamisivri@hotmail.com

© Springer International Publishing Switzerland 2018   155
A. Ustundag and E. Cevikcan, *Industry 4.0: Managing The Digital Transformation*,
Springer Series in Advanced Manufacturing, https://doi.org/10.1007/978-3-319-57870-5_9

the continuous generation of high volume, high velocity and complex data which is also known as Big Data (Lee et al. 2015a). Therefore, data analytics becomes more important for factories due to these rapid developments on the data domain.

In today's industry, how to utilize data to understand current conditions and detect faults is considered as an important research topic (Lee et al. 2015b). The use of data analytics to improve the performance of manufacturing systems has been extensively considered in the literature. In fact, data modeling and analytics are an integral part of almost any data-driven decision making. The Industry 4.0 manufacturing systems include so many sensors, interconnected devices and machines that facilitate collection of a huge amount of data. So that, manufacturing firms have many opportunities for improving the performance of their manufacturing processes. Therefore, because of Industry 4.0 trend, big data technologies are shifting from data collection to data analyses and outcome (Esmaeilian et al. 2016).

Data Analytics helps manufacturing firms to get actionable insights resulting in smarter decisions and better business outcomes (Jain 2017). For this reason, data analytics is becoming a very attractive topic for almost every manufacturing firm in Industry 4.0 era.

Generally the data analytics is covered under three sub topics. First one is *Descriptive Analytics* that summarizes the data and reports the past. It answers the question "What has happened?" and extracts information from raw data (Delen and Demirkan 2013). There is also an extension to the descriptive analytics named "diagnostic analytics" which reports the past but tries to answer the questions like "Why did it happen?" (Soltanpoor and Sellis 2016).

Second sub topic is *Predictive Analytics* which is considered as the forecasting phase. It includes the descriptive analytics output as well as some machine learning (ML) algorithms and techniques to build accurate models that predict the future. It answers the questions "What will happen?" and "Why will it happen?" in the future (Delen and Demirkan 2013).

Finally, goal of the *Prescriptive Analytics* is to provide business value through better strategic and operational decisions. It is all about providing advice. In general, prescriptive analytics is also a predictive analytics which prescribes some courses of actions and shows the likely outcome or influence of each action. It answers the questions "What should I do?" and "Why should I do it?" (Soltanpoor and Sellis 2016).

In this paper, we are going provide some examples of predictive analytics modeling that is the most interested data analytics topic in Industry 4.0. The rest of the paper is as follows. In Sect. 9.2, a literature review on Industry 4.0 and Data analytics in Manufacturing is provided. In Sect. 9.3, the techniques used in the application phase are explained. In Sect. 9.4, an application from a real world manufacturing facility is given. And finally Conclusion is given in the last section.

9.2 Literature Review

In this section, we aim to explain some manufacturing oriented data analytics case studies in the literature. The applications we focus on are applied in manufacturing area and are all include some predictive analytics techniques. The case studies are

Table 9.1 Studies about predictive analytics in manufacturing

Authors and Year	NN	SVM	Bayes	Decision trees	KNN	Linear regression
Hedge and Gray (2017)				✓		✓
Mehta et al. (in press)	✓	✓		✓	✓	
Wu et al. (2017)				✓		
Zhang et al. (2017)			✓			
Anicic et al. (2017)	✓					
Kang and Kang (2017)	✓					
Loyer et al. (2016)	✓	✓		✓		✓
Melhem et al. (2016)						✓
Lee et al. (2015b)		✓				
Shin et al. (2014)	✓					
Lee et al. (2014)			✓			
Lieber et al. (2013)				✓	✓	
Li et al. (2009)	✓					

selected to ensure that the utilize data analytics in Industry 4.0 and that they emphasize significant impact on operational effectiveness of manufacturing firms.

Some of the recent studies in the literature can be summarized as follows. Melhem et al. (2016) have tried to predict the product's quality in the semiconductor manufacturing process by using regularized linear regression models. Mehta et al. (n.d.) have compared Neural Networks, Decision Trees, Support Vector Machines and k-Nearest Neighbor methods on continuous manufacturing process. While Li et al. (2009) proposed a Fuzzy Neural Network for metal cutting industry, Wu et al. (2017) has tried to improve results by using Random Forests algorithm for the same case. Zhang et al. (2017) has provided another predictive model that includes Naïve Bayes classification technique for pharmaceutical industry. An extended Neural Network model for forecasting heat affected zone fore of the laser cutting process has been provided by Anicic et al. (2017). In addition Kang nad Kang (2017) has also used another extended Neural Network model for build an intelligent virtual metrology system for semiconductor manufacturing. Table 9.1 shows recent predictive analytics studies and the techniques used in the mentioned study.

In addition to above researches, we are going to provide some predictive analytics cases in detail with proposed frameworks.

9.2.1 Power Consumption in Manufacturing

Shin et al. (2014) present a predictive analytics model in the metal cutting industry. It is emphasized that using machine learning and predictive analytics drives manufacturing firms to make better decisions. They adopt a data-driven analytic

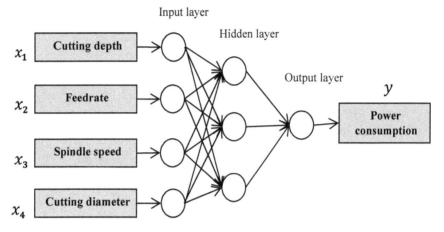

Fig. 9.1 A structure of neural network (Shin et al. 2014)

modeling approach that classify a machining operation in which manufacturing firms can be use cause (planning) and effect (monitoring) factors.

In the model, functional flow is divided into pre-production, production and post-production parts. Design of predictive analytics model, modeling and optimization are listed as post-production steps.

After the preparation of training data set, a machine-learning model is provided which involves supervised learning and analyzes the training pair sets and infers a function. A back propagation neural networks algorithm is selected to develop the unit analytic model for the learning, because the algorithm has strength for figuring out the complex relationship due to many learning input sets (Fig. 9.1).

Three unit analytic models are derived for Steel, Aluminum and Titanium in terms of a workpiece material. The machine learning system generates three predictive models for each material type from the training set.

As a result, Normalized Root Mean Square Error (NRMSE) of 'steel' marks 17%, 'aluminum' 21% and 'titanium' 11% respectively. Because of the range of the learning input data is too narrow, the accuracy is not so good. On the other hand, the method can find an applicable data pattern in unknown data correlation with well-arranged range of training data set.

9.2.2 Anomaly Detection in Air Conditioning

Lee et al. (2015b) suggest a framework for Prognostics and Health Management (PHM). System is described with a focus on three interactive agents: a System Agent (SA), a Knowledge Agent (KA), and an executive agent (EA). The SA is responsible for the management of the hardware resources, the data acquisition board and the connections. The KA is responsible for management of knowledge rules, algorithms, offline models, EA generation and reporting. Finally, responsibilities of the EA are listed as online data management, prognostic analysis execution and management of external communications.

It is emphasized that, downtime from the air compressors can cause severe losses in productivity in manufacturing factories. Lee et al. (2015b) investigate early detection of a surge on the air compressor and integrate the detection method with the control system to avoid the surge from occurring.

A lot of sensors are installed for monitoring the compressor, finally there are 20 variables measured and calculated. Three of them provided as key variables to detect the surge condition.

An asymmetric support vector machine (ASVM) classification algorithm is used to automatically determine the compressor condition. It is cited that ASVM method would not miss any detection, but would create false alarms. Of course, a missed detection is more risky and costly than a false alarm. The trained classification boundary can be used to provide early detection of the surge condition in the real time monitoring. The algorithm resulted in a 100% classification rate.

If the interactive agents are analyzed described in above framework, KA appears to have been used only once to load the trained classification boundary and model parameters for the ASVM algorithm. Works of EA can be listed as to perform the majority of the calculation by using the monitored signals and the trained classification algorithm and to provide an early detection of the surge condition. Lastly, the SA would interface with the control system to quickly avoid the surge condition if a problem was detected.

Finally, it is cited in the study that the manufacturing factory worked with the original equipment manufacturer of the air compressors to eventually embed this model into the compressor control system.

9.2.3 Smart Remote Machinery Maintenance Systems with Komatsu

At the Industry 4.0 era, factory transformation and production management are expected to reach a new level with data analytics and cyber-physical systems. For more intelligent decision making in the manufacturing process, a Cyber-Physical System framework for self-aware and self-maintenance machines has been provided by Lee et al (2014).

Within the scope of the research, physical space is considered as a fleet of machines and human actions. Machine part includes collected data, control parameters, machine performance, machine configuration, model information and task history while human part consists of maintenance activities and operational parameters.

There are three steps for the cyber (calculation) space respectively. Firstly, the data and information format are expected to be well defined for recording and management of the information collected from the physical space. Secondly, the cyber space design is expected to be able to summarize and accumulate knowledge on machine degradation that can be used for health condition of new machines. Finally, the results should be fed back to the physical space to be taken the proper action.

In the research, a mechanism in which different machine performance behaviors can be accumulated and utilized for future health assessment has been provided. Unsupervised learning algorithms are suggested for autonomously creating clusters for different working regimes and machine conditions (Fig. 9.2).

An application is also provided for predicting the health of the diesel engine of a heavy-duty equipment vehicle used in mining and construction. The parameters include pressures, fuel flow rate, temperature and the rotational speed of the engine.

For the data preprocessing step, an auto-regressing moving average approach is used to predict a time series value a few steps ahead to replace missing values.

After preprocessing the data, the next step is to develop a methodology to classify the different engine patterns in the data. Bayesian Belief Network (BBN) classification technique is used. BBN is based on the manufacturer's experience on engine related problems, along with the pattern history of the data to build the model. This classification model is able to interpret the anomalous engine behavior in the data and identify the main cause of the problem at the early stage of degradation.

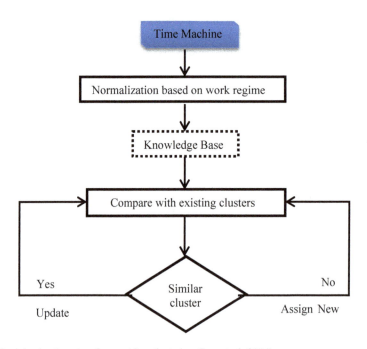

Fig. 9.2 Adaptive learning for machine clustering (Lee et al. 2014)

Finally, a fuzzy logic-based algorithm is used for the remaining life prediction. The fuzzy membership functions are based on engineering experience as well as features extracted from the data patterns. This hybrid approach accounted for the uncertainty in the data and combined data-driven and expert knowledge for a more robust approach.

9.2.4 Quality Prediction in Steel Manufacturing

In the steel industry, product quality deviations can be high due to resource-consuming, complex and automated interlinked manufacturing processes. Depending on this situation, manufacturing firms are trying to predict physical quality of products in manufacturing processes as early as possible. The rolling mill case study of Lieber et al. (2013) also explains such a problem.

In this study, after different types of sensor measurements like rolling force, speed and temperature, data are preprocessed and feature extraction procedure are implemented. For each individual steel bar quality labels such that "OK" and "NOK" are assigned. Due to the fact that, processes leading to a low product quality very close to processes leading to a high quality, end dimension of final product is used instead of quality labels.

The nearest neighbor method is used for assessing how well both classes can be separated quantitatively. Therefore, seven features that include the heating time in the hearth furnace, rolling force, speed and temperature are selected.

As a result, decision tree and k-NN methods are used with the features of the first finishing roll. Accuracy of decision tree method computed as 90%, while k-NN (k = 11) achieved about 97%. In addition, the position of the roll (channel 501) is found as the most important feature for deciding, since it determines the height of the end product (Fig. 9.3).

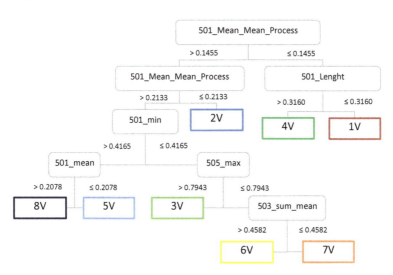

Fig. 9.3 Decision tree for detecting the final product's end dimension (Lieber et al. 2013)

9.2.5 Predicting Drilling Efficiency

Another case that used machine learning techniques in manufacturing is come across with Hegde and Gray (2017)'s work. They used regression and bootstrapping in their previous work to predict the rate of penetration (ROP) during drilling. In this case, random forests technique is preferred to predict the ROP.

Weight on bit (WOB), rotations per minute of the drill bit (RPM), flow rate of the drilling mud, rock strength (UCS) are used as input parameters in the machine learning model. On the other hand, it is cited in the article that increasing the number of relevant input parameters may yield a model with higher accuracy. Other variables such as mud properties, drill string configuration, logs and bottom-hole assemblies were not included in this work.

In the data exploration step, some outliers are removed from data. A pairs plot was used to determine the correlations between input parameters and the ROP. These correlations could be useful for model construction and selection of important features. After analyzing the plot, some input features could be discarded (Table 9.2).

As many data analytics techniques, data are divided into different sets; training and test. This operation helps to avoid overfitting. As a result, random forests technique has predicted ROP with lower error than linear regression in a given sandstone formation. R^2 for random forests is 0.96 and linear regression is 0.42. RMSE using the random forest algorithm is 7.36 ft/h, less than half of the RMSE for linear regression (18.43 ft/h). The mean error for random forests (%5) has good results in comparison to linear regression which shows a normalized error of 14%.

Finally, predicted results are used for to change surface parameters on the rig, so that ROP has been expected to increase while drilling. Also, a brute force algorithm is used to modify surface parameters to reach maximum ROP. WOB, RPM, and flow rate are optimized for the length of the well to achieve an increase in ROP.

9.2.6 Estimation of Manufacturing Cost of Jet Engine Components

In another study, Loyer et al. (2016) try to estimate the manufacturing cost of civil jet engine components during the earliest phase of the design process. Different data

Table 9.2 Correlation matrix for limestone rock (Hegde and Gray 2017)

ROP	0.80	0.47	0.082	0.39	0.83
	Depth				
0.44	0.095	0.42	0.80		
RPM	0.07	0.14	0.37		
	WOB	0.36	0.019		
		Flow rate	0.46		
			UCS		

analytics methods are provided in the paper and improvement on the forecasting accuracy are explained step by step.

The dataset contains 254 data points, 6 covariates and manufacturing cost for the year of 2012 as the output variable of the model to be explained. The observations divided unevenly among 5 different types of large engines of the top five manufacturers worldwide and 2 part categories: intermediate-pressure (IPC) and high-pressure compressor (HPC) blades. 6 primary covariates have been proposed in the case. The first one is the part category (IPC, HPC), second are related to geometry (span, chord), two to material properties (machinability, cost rate) and one to the economics of the product (accumulated production volume).

Five predictive models are applied to the estimation of manufacturing cost of mechanical components. These are Multiple Linear Regression (MLR), Artificial Neural Networks (ANN), Generalized Additive Models (GAM) which is an extension of linear regression model, Support Vector Machines (SVM) and Gradient Boosted Trees (GBT) which is a type of regression decision trees. The best one is indicated as Gradient Boosted Trees that presents a dramatic improvement over Multiple Linear Regression and Support Vector Machines in prediction accuracy. Accuracy measurement of all models have been provided and comparison of the general characteristics of the models are also has been listed in terms of goodness of fit, prediction accuracy, interpretability, easiness to fit and train, extreme values, computing affordability.

The decision tree model has given the best result with 0.96 R^2 and 6.40% MAPE while linear regression was the worst one with 0.62 R^2 and 18.07% MAPE.

If we look at the characteristics of the models, GBT and SVM are the best for goodness of fit while GBT is the best one for accuracy. As linear regression is the simplest model, interpretability and easiness has been found as best according to other models.

9.3 Methodology

The aim of this section is to summarize the techniques that are used for predictive analytic and give insight about how to compare different models used for prediction.

Predictive analytics modeling is the task of building a concept model that expresses the target (dependent) variable as a function of the explanatory (independent) variables. Minimizing the error, the difference between the predicted and real values, is considered as the goal of predictive modeling. A predictive modeling schema is represented in the Fig. 9.4. The predictive model consists of a set of parameters that are the attributes, operators and constants. Predictive modeling is the process of tuning or training these parameters of the model using a supervised learning algorithm to fit a set of instances of the concept as well as possible. The instances that are used to build the model are consequently called the training

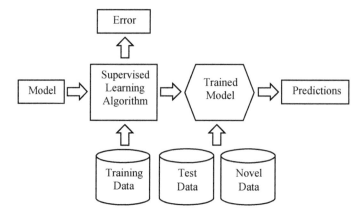

Fig. 9.4 Predictive modeling (König 2009)

dataset. A model can have a predefined static structure or it can be developed dynamically during training (König 2009).

9.3.1 Techniques Used for Predictive Analytics

9.3.1.1 Linear Regression

Linear Regression is the most known regression algorithm that aims to find the slope of the output with respect to the input. The idea also emphasizes that which of the variable how much contributes to the target variable. Each attribute is assigned a factor w_i and one extra factor is used to constitute the base level of the predicted attribute (Abbott 2014).

$$y_i = w_0 + w_1 x_1 + w_2 x_2 + \cdots + w_n x_n \tag{9.1}$$

9.3.1.2 Logistic Regression

Logistic regression is interpreted as a linear classification technique for binary classification. The ratio of the outcome probabilities is the core of the logistic regression which is named odds ratio.

$$odds\ ratio = \frac{P(1)}{1 - P(1)} = \frac{P(1)}{P(0)} = w_0 + w_1 x_1 + w_2 x_2 + \cdots + w_n x_n \tag{9.2}$$

The calculation of the probability that the outcome is equal to 1 is: (Abbott 2014)

$$\Pr(target = 1) = \frac{1}{1 + e^{-(w_0 + w_1 x_1 + w_2 x_2 + \cdots + w_n x_n)}} \qquad (9.3)$$

9.3.1.3 Support Vector Machines

The support vector machine (SVM) determines the support vectors of a hyperplane which separates the observations of two classes with maximum margin. It provides a kernel function that measures the similarity of observations in a higher dimensional feature space. So that, this issue allows for a nonlinear separation of observations in the original input space (Lieber et al. 2013).

9.3.1.4 Neural Networks

As understanding the name of the technique, *Neural Network* is a multilayer network that consists of three classes. Input layers receive information to be processed while output layers are where the results of the processing are found. Finally, hidden layers take place in between known layers (Kotsiantis 2007).

It is common to describe the structure of a neural network as a graph whose nodes are the neurons and each (directed) edge in the graph links the output of some neuron to the input of another neuron. Each neuron receives as input a weighted sum of the outputs of the neurons connected to its incoming edges. Feedforward networks are the most used type of NN in which the underlying graph does not contain cycles (Shalev-Shwartz and Ben-David 2014).

9.3.1.5 K-Nearest Neighbor

The nearest neighbor algorithm is a non-parametric algorithm that is among the simplest of algorithms available for classification (Abbott 2014). It stores a set of labeled observations. New observations are classified by majority vote of the k nearest neighbors (Lieber et al. 2013).

9.3.1.6 Decision Trees

Decision trees are among the most popular predicting modeling techniques in the analytics industry. They are easy to understand and build besides handle both nominal and continuous inputs. CHAID (Chi-square Automatic Interaction Detection), C5.0 (Quinlan algorithm), CART (Classification and Regression Trees) are most known decision tree algorithms (Abbott 2014).

Decision trees classify observations by sorting them into axis parallel rectangular regions of the input space. The method recursively determines features whose values can be used for sorting observations into regions that contain as many points of the same class as possible. The actual classification is then performed by tests on the chosen features and their values, along a path from the root to the leaves of the tree (Lieber et al. 2013).

9.3.1.7 Naïve Bayes

The Naïve Bayes methods use knowledge of probability and statistics based on applying Bayes' theorem, which can predict the class membership probabilities (Zhang et al. 2017). The parameter c indicates the positive and negative class variable, $F = (f_1, f_2, \ldots, f_n)$ stands for the object and the $(f_1, f_2, \ldots, f_n)$ represents the feature variables. P(c) is prior probability or marginal probability, P(F) is constant for all classes, P(c|F) and P(F|c) denotes the posterior probability and conditional probability, respectively.

$$P(c|F) = \frac{P(F|c)P(c)}{P(F)} \tag{9.4}$$

$$P(F|c) = \prod_i^n P(f_i|c) \tag{9.5}$$

$$f_{nb}(F) = \frac{P(c = +)}{P(c = -)} \prod_i^n \frac{P(f_i|c = +)}{P(f_i|c = -)} \tag{9.6}$$

9.3.2 Forecast Accuracy Calculation

9.3.2.1 Training and Test Sets

Forecast accuracy is measured how well the model performs on the novel data. Therefore, a portion of the available data is generally used for testing, and the rest of it is used for training the model. Then the testing data can be considered used to measure how well the model is likely to forecast on new data. The size of the test data is typically set about 20% of the total sample. While determining the training and test data the following issue should be considered: (Hyndman 2014)

- If a model fits the data well, it does not guarantee the forecast well.
- For a perfect fit, enough parameters are needed.
- As well as failing, over-fitting is also a bad situation for a model.

9.3.2.2   Forecast Accuracy Measures

The forecast errors are the difference between the actual values in the test set and the forecasts produced using only the data in the training set (Hyndman 2014).

$$e_i = y_i - \hat{y}_i \tag{9.7}$$

Scale-dependent errors
These errors are on the same scale as the data. The two most commonly used scale-dependent measures are based on the absolute errors or squared errors.

$$Mean\ absolute\ error\ (MAE) = mean(|e_i|) \tag{9.8}$$

$$Root\ mean\ squared\ error = \sqrt{mean(e_i^2)} \tag{9.9}$$

Percentage errors
Percentage errors have the advantage of being scale-independent and so are frequently used to compare forecast performance between different data set.

$$p_i = 100\frac{e_i}{y_i} \tag{9.10}$$

The most known measure is MAPE.

$$Mean\ absolute\ percentage\ error\ (MAPE) = mean(|p_i|) \tag{9.11}$$

Scaled errors
This type of errors were considered as an alternative to using percentage errors when comparing forecast accuracy across series on different scale (Hyndman and Koehler 2006). A scaled error is computed as $q_i = e_i/Q$ where Q is a scaling statistic computed on the training data. For seasonal and non-seasonal time series, the calculation of Q is different. As a result, MASE is determined as follows:

$$Mean\ absolute\ scaled\ error\ (MASE) = mean(|q_i|) = \frac{MAE}{Q} \tag{9.12}$$

9.4   A Real World Case Study

9.4.1   Definition of the Problem

Most of companies in manufacturing industry need to predict production time to ensure customer satisfaction. As an example of this situation, a shoe production company case is going to be provided in this chapter.

Main shoe production steps are: as cutting, skiving, mounting and cleaning. Every step takes different time and product type could also be differentiate the production time. In the factory, five different types of shoe which are shoe, boots, top boot, sandals and slipper are being produced. There are some varying steps to be followed while a shoe is produced according to type of product, material, gender and season. For instance, a shoe should be cut by hands while another one could be cut by the leather cutting machine. In addition, the mounting type can be changed by manually or by machine. Therefore, the company needs to predict the production time for every new product (Fig. 9.5).

9.4.2   Data Gathering and Cleaning

The first step of data analytics is considered as data collection. In the most of manufacturing cases, data is collected by sensors or networking machines. On the other hand, in some cases, data should be collected manually. In our case, data is collected using a mobile software used by the employees. Therefore, production time is collected by the employees and independent variables are obtained from the current ERP software.

The first step of data analytics is considered as data gathering. In the most of manufacturing cases, data is collected by sensors or networking machines. On the other hand, in some cases, data should be collected manually. In the type of case like mentioned above, data generally collected by workers. Therefore, the shoe company wanted its employees to note the production time and independent variables to obtain the required data of the model.

Cutting   Mounting

Skiving   Cleaning

Fig. 9.5 Shoe production steps

Table 9.3 Dataset of shoe production

Material	Gender	Season	Process type	Product type	Cutting type	Mounting type
Leather	Woman	Winter	Cutting	Shoe	Hand	Hand
Imitation	Man	Summer	Skiving	Boots	Machinery	Machinery
Textile			Mounting	Top boot	Press	
			Cleaning	Sandals		
				Slipper		

After the outlier and missing data elimination, a dataset that include seven independent variable and production time of 57,600 shoes was obtained. The independent variables and their values are provided in Table 9.3. Production time is noted in seconds in dataset.

9.4.3 Model Application and Comparisons

After the data gathering and cleaning step, data are divided into training and test datasets. Decision Tree, K-nearest neighborhood (k-NN) and artificial neural networks (ANN) techniques are used to train models and the train models are applied to the test data set. The accuracy of the models are compared using Mean Absolute Percentage Error (MAPE) and Mean Absolute Error (MAE) metrics. According to results, best performance was obtained with CHAID decision tree model and the most important variable was found as the process type. This model is a classification method for building decision trees by using chi-square statistics to identify optimal splits and divides production times into bins as a dependent variable that differentiate respect to independent variables.

The results reveal that Decision Tree model gives the best prediction for the above mentioned problem. Table 9.4 shows the results of other models besides the decision tree model (Fig. 9.6).

Table 9.4 Comparison of predictive algorithms

Model	MAPE (%)	MAE
CHAID	3.20	5.64
k-NN	3.47	6.16
Neural network	6.77	7.25

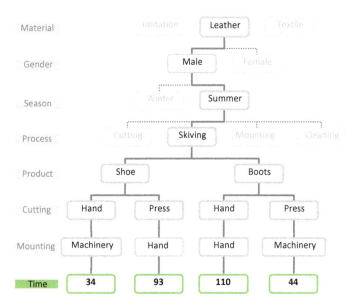

Fig. 9.6 A part of resulting decision tree for shoe production

9.5 Conclusion

With the technological advances and the intensive use of sensors, manufacturers
have to deal with more intensive and more complex data. This new case makes data
analytics more important for the factories. In the Industry 4.0 era, data analytics
focuses predictive analytics rather than descriptive analytics. Predictive analytics
tries to build a model that predicts the target variable from set of independent
variables. In the model, many of supervised algorithms including classification and
regression algorithms could be used. Error metrics are used to determine which
algorithm yields better results and the performance of the model.

Current studies in the literature shows that predictive analytics is used in many
points of manufacturing. As a real case study provided in this paper, manufacturing
firm tries to predict the production time of different type of products and processes.
After data preprocessing step, supervised learning methods are used to find best
result for the problem. CHAID decision tree technique gives the best result for the
shoe production case.

As a continuous learning process, manufacturing firm can be create a dynamic
predictive model that update itself according to changes in dataset. Also, changes in
human and machinery factors can appended to the dataset. With use of extended
techniques that come with the evolving use of machine learning algorithms can give
better results for the manufacturing analytics cases.

Acknowledgements This work is supported by Scientific and Technological Research Council of
Turkey (TUBİTAK), TEYDEB 1507, Grant No: 7141451.

References

Abbott D (2014) Chapter 8—Predictive modeling. Applied predictive analytics: principles and techniques for the professional data analyst. Wiley, Hoboken, pp 213–281

Anicic O, Jović S, Skrijelj H, Nedić B (2017) Prediction of laser cutting heat affected zone by extreme learning machine. Opt Lasers Eng 88:1–4

Delen D, Demirkan H (2013) Data, information and analytics as services. Decis Support Syst 55(1):359–363

Esmaeilian B, Behdad S, Wang B (2016) The evolution and future of manufacturing: a review. J Manuf Syst 39:79–100

Hegde C, Gray KE (2017) Use of machine learning and data analytics to increase drilling efficiency for nearby wells. J Nat Gas Sci Eng 40:327–335

Hyndman R (2014) Chapter 8—Forecasting performance evaluation and reporting. Business forecasting: practical problems and solutions. SAS Institute Inc., pp 177–184

Hyndman RJ, Koehler AB (2006) Another look at measures of forecast accuracy. Int J Forecast 22:679–688

Jain VK (2017). Chapter 1—Overview of big data. Big data and Hadoop. Khanna Book Publishing Co Ltd

Kang S, Kang P (2017) An intelligent virtual metrology system with adaptive update for semiconductor manufacturing. J Process Control 52:66–74

Kotsiantis SB (2007) Supervised machine learning: a review of classification techniques. Informatica 31:249–268

König R (2009) Predictive techniques and methods for decision support in situations with poor data quality. University of Boras, School of Business and Informatics, University of Skovde, informatics Research Center, University of Orebro, School of Science and Technology. Örebro University, 112 p

Lee J, Bagheri B, Kao H-A (2015a) A cyber-physical systems architecture for industry 4.0-based manufacturing systems. Manuf Lett 3:18–23

Lee J, Kao H, Ardakani HD, Siegel D (2015b) Chapter 19—Intelligent factory agents with predictive analytics for asset management. In: Industrial agents. Elsevier Inc., pp 341–360

Lee J, Kao H-A, Yang S (2014) Service innovation and smart analytics for industry 4.0 and big data environment. Procedia CIRP 16:3–8

Li X, Lim BS, Zhou JH, Huang S, Phua SJ, Shaw KC, Er MJ (2009) Fuzzy neural network modelling for tool wear estimation in dry milling operation. In: Annual conference of the prognostics and heath management society, pp 1–11

Lieber D, Stolpe M, Konrad B, Deuse J, Morik K (2013) Quality prediction in interlinked manufacturing processes based on supervised & unsupervised machine learning. Procedia CIRP 7:193–198

Loyer J-L, Henriques E, Fontul M, Wiseall S (2016) Comparison of machine learning methods applied to the estimation of manufacturing cost of jet engine components. Int J Prod Econ 178:109–119

Mehta P, Butkewitsch-choze S, Seaman C (in press) Data analytics framework for semi-continuous manufacturing process—implementation vision with a use case. J Manuf Syst

Melhem M, Ananou B, Ouladsine M, Pinaton J (2016) Regression methods for predicting the product's quality in the semiconductor manufacturing process. IFAC-Papers OnLine 49(12):83–88

Qin J, Liu Y, Grosvenor R (2016) A categorical framework of manufacturing for industry 4.0 and beyond. Procedia CIRP 52:173–178

Shalev-Shwartz S, Ben-David S (2014) Chapter 20—Neural networks. Understanding machine learning : from theory to algorithms. Cambridge University Press, Cambridge, pp 269–282

Shin S-J, Woo J, Rachuri S (2014) Predictive analytics model for power consumption in manufacturing. Procedia CIRP 15:153–158

Soltanpoor R, Sellis T (2016) Prescriptive analytics for big data. In: Cheema MA, Zhang W, Chang L (eds) Databases theory and applications. Paper presented at the 27th Australasian Database Conference: ADC 2016. Springer International Publishing, Sydney, NSW, pp 245–256

Wu D, Liu S, Zhang L, Terpenny J, Gao RX, Kurfess T, Guzzo JA (2017) A fog computing-based framework for process monitoring and prognosis in cyber-manufacturing. J Manuf Syst 43:25–34

Zhang H, Kang Y, Zhu Y, Zhao K, Liang J, Ding L (2017) Toxicology in Vitro Novel naïve Bayes classification models for predicting the chemical Ames mutagenicity. Toxicol In Vitro 41:56–63

Chapter 10
Internet of Things and New Value Proposition

Gaye Karacay and Burak Aydın

Abstract Internet of Things (IoTs) are the new wave in technological innovation that fundamentally shift the dynamics of businesses all around the world. Contrary to popular belief, IoTs are not just about sensors or machine intelligence embedded in variety of things that are part of our business or personal lives, but these are the tools that provide the primary base for doing business in a novel and integrated way by "creating value" as social, personal, and economic return. Value creation triggered by IoT via provision of variety of data is the core of these novel business systems. Gradually every industry will be disrupted by the emergence of new data which become easily available by means of IoTs and their various applications. These novel technologies will bring about not only new hardwares, new applications, and new services; but also will fundamentally change the processes or the ways of doing work. This chapter looks at how IoTs enable creation of new value in business life by examining real life IoT applications within different sectors.

10.1 Introduction

The Internet era, started at the second half of the twentieth century, introduced a whole new paradigm of communication caused by switching from analog to digital technologies, and bringing in internet and World Wide Web (www). The digital revolution along with the amplified globalization opened the doors for the Third Industrial Revolution, and had its latest achievement by the creation of social networks, such as Facebook, Twitter, and LinkedIn. Moreover, a new breed of

G. Karacay (✉)
Department of Industrial Engineering, Faculty of Management, Istanbul Technical University, 34367 Macka, Istanbul, Turkey
e-mail: karacayaydin@itu.edu.tr

B. Aydın
Silver Spring Networks, Europe Middle East and Africa (EMEA), San Jose, US

© Springer International Publishing Switzerland 2018   173
A. Ustundag and E. Cevikcan, *Industry 4.0: Managing The Digital Transformation*,
Springer Series in Advanced Manufacturing, https://doi.org/10.1007/978-3-319-57870-5_10

low-cost communication technologies have made it possible for everyday objects to be part of these social networks via novel devices. Meanwhile, on-going advancements in sensor and radio-frequency identification (RFID) technologies make real-time data tracking ever much easier, and so enable constant real time data exchange between machines. These advancements engender creation of systems made up of machines, materials, or things which are capable of communicating autonomously with each other along the value chain activities by creating huge amounts of data available for further analysis, all referred to as *Internet of Things* (IoTs). Further increase in the amount of real time data exchange between these *'things'*, together with the expansion of digitalization in the manufacturing industry, and the accelerated progress in sophisticated cyber-physical devices have jointly changed the nature of supply and demand markets in various industries. The innovations in digital technology have changed the supply side of the market by making the sources of supply more accessible and economic, while they have transformed the demand side of the market by shifting customer demand by providing the undistorted information about markets and products. As a result, novel business models have emerged by forcing industries to redefine their value propositions which have disturbed market dynamics as well as the dominant positions of incumbent companies. In consequence of these technological advancements and business transformations, the Fourth Industrial Revolution has started at the beginning of the twenty first century.

While the first phase of the Forth Industrial Revolution during 1995–2005 time period were shaped by consecutive web page developments and associated corporate applications, the second phase that occurred from 2005 to present is the era where we have witnessed to an intense market exploitation of smart phones and mobile applications. From now on, we will live through to the third wave of the Fourth industrial revolution which is likely to be sort of a tsunami by taking into account the current magnitude of disruptions within various industries and businesses. In this vein, prospective change may be massive through extensive transformations of business processes and developments of entirely novel business models, yet all signal for an amplified value creation potential via IoTs based systems and their various business applications.

In today's business environment, it is fair to conclude that technology doesn't just shape strategy, it fundamentally defines it. In this respect, applications of IoTs have been fundamentally transforming how organizations create value. As stated by Deloitte Digital *"The IoT has implications not only for consumer-facing offerings, but it can also reveal profound insights into business-to-business flows across the supply chain, providing far deeper and more nuanced insight into how capital is being deployed"*. In view of that, IoTs will transform tech-ready industries in a rapid pace while it will also have significant as well as more transformative effect on industries that presently have not been technology-based yet.

10.2   Internet of Things (IoTs)

What is *Internet of Things (IoTs)*, and why it is important for businesses of today and future? IoTs, indeed, is used as a catchphrase for automated and connected things within different components of economy. Although various parties may name it differently, i.e. *Internet of Everythings*, or *Internet of Your Things*, etc., all these concepts are referred to indicate a similar meaning, which is stated by the open source dictionary Wikipedia as follows, "*The Internet of things (IoT) is the inter-networking of physical devices, vehicles (also referred to as "connected devices" and "smart devices"), buildings, and other items—embedded with electronics, software, sensors, actuators, and network connectivity that enable these objects to collect and exchange data.*" IoTs describe a network of internet-connected devices that are able to collect and exchange data using their embedded sensors. The concept revolves around networks of data-gathering sensors from very different resources ranging from watches, autonomous cars, and thermostats to manufacturing facilities that process at edge or cloud depending on the business model; and in this way create value for the user, either corporate or consumer, and usually both. Figure 10.1 shows a basic IoT Ecosystem with different layers. In every layer, a mixture of hardware, software and service components create a part of the value chain.

By the proliferation of connected sensors, which create a digital world that is progressively more quantifiable and accessible, data obtained from autonomously interconnected physical devices enable the translation of a physical world into a digital one. By means of analyzing the '*big data*' obtained from such a huge digitalized world, understanding the current trends in various markets and structures have become possible, and so, companies, economies, or countries can make further projections about various markets, as well as personal and public sector activities. Briefly, that explains how IoTs create value for businesses and overall economy.

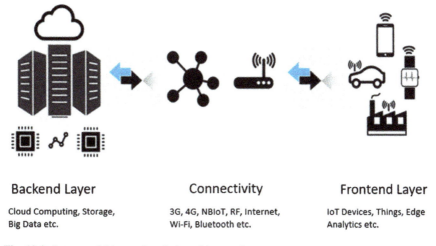

Backend Layer	Connectivity	Frontend Layer
Cloud Computing, Storage, Big Data etc.	3G, 4G, NBIoT, RF, Internet, Wi-Fi, Bluetooth etc.	IoT Devices, Things, Edge Analytics etc.

Fig. 10.1 Internet of things value chain and keywords

Generally, there are four market drivers which have amplified the integration of IoTs within current economic systems, these as:

- *Increased internet penetration*: According to International Telecommunication Union, World Telecommunication/ICT Development Report and Database, as well as the World Bank estimates, by 2015 the penetration rate of world internet users is over 40%. In majority of the developed countries this rate goes over 70% (e.g., Finland 92.7%, UK 92%, EU Average 79.6%), and in the emerging markets this rate varies between 15% and 70%, while OECD average is 77.2%. Considering that the world average of the penetration rate of internet users was 29.15% by 2010, there has been a significant growth took place between 2010 and 2015.
- *Increased mobile adoption*: Similarly, according to International Telecommunication Union, World Telecommunication/ICT Development Report and Database, and the World Bank estimates, worldwide average of the mobile cellular subscription rate jumped from 76.5% in 2010 to 98.6% in 2015.
- *Low-cost sensors*: According to World Semiconductor Trade Statistics there has been a significant drop occurring in sensor prices. For example, accelerometers which are used in smartphones and video consoles were priced as 14 USD in 2007, while they were sold to 0.50 USD in 2015. Likewise, LIDAR, a sensor that gives an autonomous vehicle its "eyes" was costing more than twice of the vehicle itself back in 2007 by having a price of 75,000 USD, whereas it was sold only by 250 USD in 2016; and it is expected to be priced under 100 USD in five years time. When sensors were expensive, we were using them sparingly. However, as prices drop, we have started to include them within ever-increasing number of devices across a range of applications.
- *Continued Moore's Law in CPUs*: Moore's Law indicates that the number of transistors per square inch on integrated circuits doubles every two year, with a higher performance, lower cost and smaller circuits. Moore's proposition is likely to continue for CPUs at present, and the developments in the shrinking size and spaces of CPUs create momentum not only in final products like smartphones, but also in big data analytics, cloud computing, etc.

Although the emerging IoTs technologies have been experiencing a burst of activity and creativity via super excited startups and venture capital firms, yet they are still in their early stages of emergence. According to 2015 McKinsey Global Institute report, the potential impact of cross sectional IoTs applications are expected to create value of $3.9 trillion to $11.1 trillion per year in 2025 (Manyika et al. 2015). Likewise, a market analysis conducted by Boston Consulting Group (BCG) depicts that by 2020 $267 billion will be spent on IoTs technologies, products, and services. The IoTs market components with the highest revenue generation potentials are IoTs applications, and investments for services in support of IoTs based systems (Manyika et al. 2015).

The value creation caused by IoTs is not limited to one or two industries, indeed this value generation occurs in a variety of sector even intersectorally. For instance, while Industry 4.0 is the application of IoTs in the manufacturing sector, wearable technologies is its reflection of its usage in the clothing industry. A current report by McKinsey Global Institute titled "The Internet of Things: Mapping the Value Beyond Hype" provides a cross sector view of the potential impact of IoTs and its applications to be $3.9 trillion to $11.1 trillion per year in 2025. Such that, factory operations and equipment optimizations are expected to create a value around $1.2 trillion to $3.7 trillion, while IoTs applications in cities (i.e., *smart cities*), public health and public transportation have a value generation potential of $930 billion to 1.7 trillion. BCG's industry based projections point that 50% of IoT spending expected to be driven solely by manufacturing, utilities, transportation and logistics industries in 2020.

10.3 Examples for IoTs Value Creation in Different Industries

There are different ways of creating value by IoTs applications, which are mostly shaped by the industry specific dynamics. Below you will find different examples depicted from business context that illustrate how IoTs based systems and their applications are changing the ways of doing business. Note that, it is all about capturing the data that we had been aware of, but could not collect due the incapabilities in the past. By utilizing technologically more sophisticated hardware and software products, and subsequently turning data into value by making use of various applications have initiated the emergence of this new value creation.

10.3.1 Smart Agriculture

There are various examples about how IoTs transform different verticals and create business value. However, 'smart agriculture' is actually a good example given that agriculture has a direct impact on human life via contributing to the satisfaction of the most essential human need, i.e. need to eat, so that enable the survival of human kind. However, what's more interesting is that agricultural sector has a great potential for utilizing IoTs and transform the ways of doing work via IoTs applications.

There had been no significant progress witnessed in agriculture before industrial revolution, but even after the industrial revolution progress was limited by the usage of grain elevators, chemical fertilizers, and the first gas-powered tractor etc. After having a steady state period for 20 years, the methods of farming have just changed dramatically by the integration of IoTs applications. According to the UN Food and Agriculture Organization (FAO), the world will need to produce 70% more food in 2050 than it did in 2006 in order to feed the growing population of the Earth. The agriculture industry must overcome increasing water shortages, limited availability of lands, difficult to manage costs, while meeting the increasing demand and consumption. That's why agricultural production all around the world is stepping towards a new system design by means of 'smart agriculture'. Smart agriculture is composed of many different technological implementations. These applications are replacing the tough, unreliable and time-consuming traditional farming techniques with efficient, reliable and sustainable smart agriculture. By that, IoTs and their applications are taking the farming industry to the next level.

IoTs in smart agriculture include sensors that help farmers monitor soil humidity, minerals, and adjust necessary watering for getting the maximum yield. On the other hand, context aware computing helps farmers to predict the best time to harvest in line with the meteorological changes or it can help farmers to utilize cold chain technologies to transport agricultural products by the most healthy and economical way. As a concrete example, John Deere which a leading manufacturer of farming equipments has begun connecting its tractors to the internet and has created a method to display data about farmers' crop yields. BI Intelligence which is Business Insider's premium research service, predicts that IoT device installations in the agriculture world will increase from 30 million in 2015 to 75 million in 2020, by a compound annual growth rate of 20%. Similarly, OnFarm, which has created a connected farm IoT platform, expects the average farm to generate an average of 4.1 million data points per day in 2050, up from 190,000 in 2014. The U.S. currently leads the world in IoT smart agriculture, as it produces 7340 kg of cereal (e.g. wheat, rice, maize, barley, etc.) per hectare (2.5 acres) of farmland, compared to the global average of 3851 kg of cereal per hectare (Meola 2016).

According to the PwC report released at 2016, global market for commercial applications of drone technology valued over $127 billion while drone-powered solutions in agriculture was $32.4 billion (Mazur 2016). In order to understand how such a huge amount of value generation could become possible by drone-powered solutions, it is better to review possible ways of aerial and ground-based drone usages throughout the crop cycle in agriculture. In that, the following six alternative usages can be given as examples (Mazur 2016):

1. *Soil and Field analysis*: Drones can produce 3-D maps for field analysis, useful in planning seed planting patterns.
2. *Planting*: Drone-planting systems shoot pods with seeds, plant nutrients and chemicals into the soil. Different start-ups claim that drones could plant over 500 seeds per hour, compared with farmers planting about 800 seeds per day.
3. *Crop Monitoring*: Drones provide the precise development of a crop and reveal production inefficiencies, enabling better crop management in the lifecycle.
4. *Spraying*: Drones can scan the ground and spray the correct amount of liquid by utilizing different sensors that lead to increased efficiency while reducing the amount of chemicals penetrating into groundwater. It is estimated that drone spraying can be completed up to five times faster than the spraying by traditional machinery. Drone spraying controlled by sensors is more efficient than mess-agricultural aircraft spraying also.
5. *Irrigation*: Drones equipped with different sensors (i.e. multispectral or thermal sensors) can identify soil needs depending on visual computing. This computing

also allows the calculation of the vegetation index that gives information about health of the crop.

6. *Health assessment*: Drone-carried devices can identify bacterial or fungal infections of trees by scanning a crop using both visible and near-infrared lights, and analyzing multispectral images that track changes in plants and indicate their health. A rapid response can save an entire orchard.

All of the above mentioned drone technologies are a part of *smart agriculture* practices that are coupled with sensors, visual and context aware computing technologies that gather various different information to create a business value in agriculture. Country comparisons can help to visualize business value creation by smart agriculture; such that, Netherlands is a country that makes use of the latest technological innovations including IoTs technologies and employs approximately 900,000 farmers and exports $90 billion of agricultural products, whereas Turkey employs 17 million farmers and makes an export of $20 billion by utilizing very traditional agricultural methods.

10.3.2 Smart City

A study by the McKinsey Global Institute suggests that the world's 600 fastest growing cities will account for 60% of global economic growth between 2010 and 2025 (Dobbs et al. 2011). In today's modern life, cities consume most of the global energy, account for the consumption of the great majority of the world's natural resources, and output most of the world's carbon emissions. So, the need of smooth operation of urban areas by using the latest technologies is a necessity rather than a luxury.

According to a global research firm Gartner, smart cities will host nearly 10 billion IoT devices by 2020, with more vendors than ever exploring new business opportunities in smart city development, ranging from home/building security, streetlights, transportation and healthcare. These 10 billion IoT devices in cities will create value globally by means of IoTs transportation applications worth more than $800 billion per year, public health savings valued nearly $700 billion per year, and smart meters reducing the loss of electricity in distribution channels together with sensors detecting water leaks worth as much as $69 billion per year. Overall, worth of IoT applications in the cities could reach an economic impact of $930 billion to $1.6 trillion per year in 2025 (Manyika et al. 2015).

Creation of value in smart cities can be analyzed practically from two different perspectives; one from the public sector, and the other from the citizens' perspective. Through smart city formations public sector can serve its citizens by a lower cost due to efficiency increases in services via digitalization. However, it's not only the cost savings that matters, but also the increased service quality as well as an uninterrupted 24/7 service time span makes the difference regarding smart city value creation from public sector perspective. Another value creation through smart city formations comes from the presence of the quality of life indicators, such as, having the necessary indicators for the carbon emissions and methane gas levels, as well as water sensors. In parallel, managing renewable resources by means of IoTs solutions also contribute to the value creation from public sector perspective. On the other hand, from citizens perspective, value creation through IoTs applications in cities would help them having savings both in terms of time and money. For instance, smart parking solutions can help drivers save not only time but also money by saving their time for working, otherwise which would be consumed at traffic.

10.3.3 **Smart Life—Wearable Technologies**

Last decade, technology became the integral part of our social and personal lives, however recently together with the transformations in IoTs technologies and sensors, it is getting smarter and more personal via daily usages of different devices with sensors like smart watches and shoes. By early 2016 a market projection on the future of wearable technologies by a market research company CCS Insight indicated that 411 million smart wearable devices, worth $34 billion, will be sold in 2020. It is expected that wrist-based devices, like smart watches and fitness trackers will continue to dominate the wearable technology market. Smart Glass is expected to constitute 25% of the total market by 2020.

Using technology as a wearable component has initiated by the same logic that is the case in other industries; creating a value, either business or personal, and most of the time both. For example safety bracelets which supply location based information creates a solid and significant personal value. Various types of wearable panic buttons, which can be worn as bracelets, key chains, or by any other kind of accessory provide safety solutions especially for kids and women. By various applications generated for different emergency scenarios styled in different forms and usage models, a range of various wearable IoTs can be generated for personal usage.

Another striking example can be given from sports, for example, a tennis player may no longer value the tennis racquet just for the frame's stiffness, string's tension, and its weight and balance, all representing the traditional assessment points for a good racquet; but by smarter technology integrated within the tennis racquet now on a tennis player may also utilize the tennis racquet as a valuable source of information about one's tennis stroke as well as how to improve personal performance and success (Crisp 2015).

Wearable technologies are not limited for personal use only, for instance smart glasses provide business value in logistics sector by helping to find locations of specific packages by supplying details necessary for courier services. Another example can be given from healthcare sector, that is, there are various usage models of smart glasses in clinical and surgical applications in Healthcare industry. For medical professionals to adopt smart glasses in their practices, these products should be tailored to fit the various needs of medical and surgical sub-specialties. (Mitrasinovic et al. 2015). This argument can be also reflected for other verticals, such that, to facilitate a smart glass or any smart gadget being used in any vertical, it needs to be customized to fit the needs of that specific vertical.

10.3.4   Smart Health

IoTs and its various applications are also redesigning contemporary health care systems and their services by offering optimistic technological, economic, and social prospects. Indeed, medical care and health care represent one of the most attractive application areas for the IoTs (Pang 2013). IoTs redefine how apps, devices and people interact and connect with each other in delivering and receiving healthcare solutions. Remote health monitoring, fitness programs, controlling chronic diseases, and elderly care are among the various IoTs based medical applications in healthcare. Nowadays, healthcare providers have been redesigning their homecare services by utilizing IoTs applications for having an improved compliance with treatment and medication given at home.

Therefore, various digital medical instruments, sensors, and digital imaging devices can be viewed as "smart devices" or "things" constituting a core part of IoT-based healthcare services. These applications are expected to reduce costs in healthcare industry while also upgrade patients' quality of life. According to industry experts, IoTs applications in healthcare market is projected to grow from USD 41.22 billion in 2017 to USD 158.07 billion in 2022 by taking advantage of the following IoTs effects (Patel 2017):

1. Decreased Costs: Remote patient monitoring can be done on a real time basis by utilizing connectivity of the healthcare solutions, thus significantly cutting down on unnecessary visits by doctors. Most importantly, through advanced home care facilities, hospital stays and re-admissions would be decreased, thus these would help to cut down the costs.
2. Improved Outcomes of Treatment: Availability of real time patient information through connectivity of remote health care solutions via cloud computing or other virtual infrastructure provide caregivers the ability to make informed decisions based on solid evidences. By that, healthcare service can be provided on a timely manner which would possibly improve treatment outcomes.
3. Improved Disease Management: Accessing real time and continuous data of patients provides health care providers better information about disease management and help them make predictive or preemptive actions to control diseases.
4. Reduced Errors, Waste and Costs: Accurate collection of data in the process flow enable reducing system costs and wastes as well as and minimizing human-based system errors.

5. Enhanced Patient Experience: Increased accuracy rates in treatments with timely health service would likely to improve patient experiences.
6. Enhanced Management of Drugs: IoTs help to better manage of drugs and so decrease the level of a major expense item in the healthcare industry.

10.4 IoTs Value Creation Barriers: Standards, Security and Privacy Concerns

While IoTs are fundamentally changing business models and improving industry outcomes, there are some barriers that slowdown the penetration of them in value creation processes within different industries. Three major barriers in scaling IoTs adoptions in business are standards, privacy and security concerns (Fig. 10.2).

While IoTs usage within business processes have been becoming more popular, IoT devices have also increasingly become an attractive subject matter for cyber-criminal targets. More than half of all global IoTs attacks originate from China and the US. There is an exponential potential of these attacks given that the more connected devices, the more attack possibilities by hackers. According to a global cyber-security firm Symantec, 2015 was a trend record year for IoTs attacks by the emergence of eight new malware families. More importantly, it is not only IoT devices which are attacked, but also they are used to become another source for attack to other systems by means of distributed denial of service (DDoS) attacks.

AT&T surveyed more than 5000 enterprises around the world and found that 85% of enterprises are considering, exploring or implementing an IoT strategy. On the other hand according to the same survey 88% of these organizations lack full confidence in the security of their business partners' connected devices. According to a report by a global embedded systems firm WindRiver, the main reason for attacks to IoTs devices is the fact these systems either lack to get firmware updates or left with the original set-up with default parameters. Therefore, they are relatively easy targets for hackers. However, at present it is neither easy nor often practical to integrate software updating procedures for every IoTs application. For example, embedded programmable logic controllers (PLCs) used extensively in factory

Fig. 10.2 Three barriers for scaling IoT business

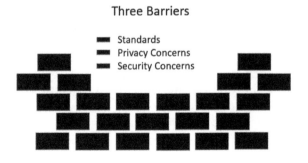

automations like robotic systems, are typically integrated as a part of corporate IT infrastructures, and the problem of these systems derive from not having a proper structure to receive software updates or security patches in a timely manner without impairing functional safety or incurring significant recertification costs every time a patch is rolled out.

10.4.1   Privacy Concerns

Privacy concerns are another huge barrier in adoptions of IoTs. Many IoTs or smart devices transmit information across networks without encryption, which creates huge risks of privacy. As an example, you may think about the smart thermostats at your home; in that, how can you be sure that you would be the only person that monitors these smart gadgets at your home, but not be thiefs, and how companies will assure the authorization of privacy for these services. Likewise, usage of smart glasses brings a specific question to our minds, that is, how your privacy will be assured while others video-record your view without your knowledge or personal consent by their smart glasses? Another striking question may be related to IoTs usage in healthcare services, explicitly, what if hackers can gather your personal health data from the smartwatch apps or healthcare tracker devices that you are using, what will be the possible outcomes of these kinds of misconduct for individuals regarding their personal safety. What are the rules and regulations, and more importantly what they should specifically be covering for? These are some examples for probable questions, and there is a significant need for everybody to develop respective answers in order find solutions for these probable drawbacks of IoTs usages.

10.4.2   Standardization

Standardization is another important barrier for IoTs adoption, and has been holding back the value generation potential of the market. Although there has been attempts coming from different stakeholders in the market for developing common standards of IoTs applications, just recently there comes a consolidating effort by some stakeholders to agree on common standards, which help IoTs market to accelerate. For example, a conglomeration of leading technology firms, including ARM, Huawei, Philips, and Vodafone have founded the Internet of Things Security Foundation whose mission is stated as "*to help secure the Internet of Things, in order to aid its adoption and maximize its benefits.*"

10.5   Conclusion

IoTs, more than any other technology ever been, have become a remarkable source of inspiration for entrepreneurs in today's world. The wide scope of innovations founded on IoTs as well as amplified digitalization across various industries have triggered the popularity of IoTs, which can be witnessed by the major business model preferences of the majority of recent startups. According to Gartner, vendors of IoTs will earn more than $309 billion in 2020 (Stevenson 2016). Indeed, what makes IoTs that much valuable for economy is their capability of data collection generated by the connected devices together with their capability to turn these data to business values by making more accurate future projections based on current trends in business and economies (Hui 2014).

Over the next 5–10 years, the number of machines and sensors connected to the internet will expected to explode, which will lead to emergence of new usage and business models. As the adoption of IoTs based systems spread, companies will fundamentally need to rethink their accepted and traditional views about value creation strategies in markets. In this vein, IoTs are not just about new smart devices, but more significantly they mean productivity improvements as well as uninterrupted connectivity of economies all over the world.

The increased demand for IoTs has roots in current shifts happening in various industries and businesses that can be expressed by the following two basic questions: How businesses can create value, and how they can capture it in today's economical dynamics shaped by changes both happening in the demand and supply side of the markets. Value creation can be in the form of hardware development, software development, services or business model innovations in IoT space. In E2E, business related value is critical for IoT applications in order to get data traction in a specific industry. Whereas for the applications of smart life or wearable technologies, IoTs space design or industry specific features leading to personal value become what really matters. Although for companies the main formula for competitive strategies still depends on having competitive positions and competitive advantages within their industries, yet in today's new market dynamics this formula need to be reshaped by the information provided by smart "things" that realign both demand and supply side of markets in various industries (Raynor and Cotteleer 2015).

According to a market analysis conducted by BCG, in 2020 almost $267 billion will be spent on IoTs technologies, and related products, and services. Two main sources of the revenue growth in IoTs market is expected to be IoT applications and associated investment for services (Manyika et al. 2015). BCG also predicts revenue from all layers of the IoT technology stack to have at least a 20% compound annual growth rate (CAGR) between 2015 and 2020. Predictive maintenance, self-optimizing production, and automated inventory management are the three top usages driving such a huge IoT market growth (Hunke et al. 2017).

However, there are also some roadblocks that slows down the adoption of IoTs technologies. Inadequate security will be a critical barrier to large-scale deployment of IoTs, and also interoperability will be one of the factors limiting this scale in near future. Privacy and legal standards also holding back possible market growth of IoTs. For this reason, it becomes significantly important for different stakeholders in IoTs market to come together and work for identifying and clearing the blocking factors in IoTs market penetration. Companies just started to explore what the IoT means for them. While some changes brought by these IoT technologies is incremental and relatively easy to adopt, yet some others are more transformative so that they would require a willingness to question some deeply held assumptions about traditional way of doing business. Proven by past examples, companies who surf over the tsunami wave will likely to thrive.

References

Crisp S (2015) Rafael Nadal demonstrates Babolat Play & Connect interactive tennis racquet," gizmag, http://www.gizmag.com/rafael-nadal-demonstrates-babolat-play–connect-interactive-tennis-racquet/22699/. Accessed 28 Feb 2015

Dobbs R et al (2011) Urban world: mapping the economic power of cities, McKinsey Global Institute, March 2011

Hui G (2014) How the internet of things changes business models, HBR, 29 July 2014

Hunke N, Yusuf Z, Rüßmann M, Schmieg F, Bhatia A, Kalra N (2017) Winning in IoT: it's all about the business processes, BCG. Perspectives, 05 Jan 2017. https://www.bcgperspectives.com/content/articles/hardware-software-energy-environment-winning-in-iot-all-about-winning-processes/

Manyika J, Chui M, Bisson P, Woetzel J, Dobbs R, Bughin J, Aharon D (2015) The internet of things: mapping the value be-yond hype, McKinsey Global Institute, June 2015. http://www.mckinsey.com/~/media/McKinsey/Business%20Functions/McKinsey%20Digital/Our%20Insights/The%20Internet%20of%20Things%20The%20value%20of%20digitizing%20the%20physical%20world/Unlocking_the_potential_of_the_Internet_of_Things_Executive_summary.ashx

Mazur M (2016) Six ways drones are revolutionizing agriculture, PWC, 20 July 2016

Meola A (2016) Why IoT, big data & smart farming are the future of agriculture, Business Insider, 20 Dec 2016

Mitrasinovic S, Camacho E, Trivedi N, Logan J, Campbell C, Zilinyi R, Lieber B, Bruce E, Taylor B, Martineau D, Dumont E, Appelboom G, Connolly S (2015) Technol Health Care 23 (4):381–401

Pang Z (2013) Technologies and architectures of the internet of-things (IoT) for health and well-being, PhD Thesis in Electronic and Computer Systems, KTH—Royal Institute of Technology, Stockholm, Sweden, January 2013

Patel K (2017) 6 Benefits of IoT for hospitals and healthcare. Retrieved from: https://www.ibm.com/blogs. Accessed on 9 Jan 2017

Raynor ME, Cotteleer M (2015) The more things change: value creation, value capture, and the internet of things, Deloitte Review, Issue 17, 27 July 2015

Stevenson A (2016) Delivering business value with internet of thing, Express Computer on 28 Jan 2016

World Bank (2017) Retrieved from: http://data.worldbank.org/indicator/IT.NET.USER.P2. Accessed on 26 April 2017

Chapter 11
Advances in Robotics in the Era of Industry 4.0

Barış Bayram and Gökhan İnce

Abstract The industrial robots in factories have been recently designed and utilized to handle dangerous tasks for humans, to achieve faster and more accurate production processes, and to reduce the cost of the products. Since the competitiveness in today's business environment increases, manufacturers require more intelligent systems making smarter decisions. In the light of Industry 4.0 revolution, the advances in information technology like artificial intelligence, cloud and Big Data change the use and design of robots in the industry. The potential industrial robotic applications and the next generation of robotics planned to be utilized in the Industry 4.0 factories are discussed.

11.1 Introduction

In factories, human workers are not effective any more on many reasons such as their capabilities and physical capacities affecting production performance, production cost, etc. Therefore, industrial robots which are machines with automated and embedded intelligence and capabilities are required to improve the manufacturing process. To achieve more accurate production at shorter time without any injuries in recent competitive industry have shifted the manufacturers' mind from human labour to robots. Nevertheless, a collaborative work of humans and robots is required for efficient and robust manufacturing.

In the modern competitive industry, companies need to facilitate industrial robots not only due to safety reasons to reduce labour force injuries during production, but also due to the need of faster and more accurate production considering economical gains. However, the recent robotic technology does not provide

B. Bayram (✉) · G. İnce
Department of Computer Engineering, Faculty of Computer and Informatics Engineering,
Istanbul Technical University, 34469 Maslak, Istanbul, Turkey
e-mail: baris.bayram@itu.edu.tr

G. İnce
e-mail: gokhan.ince@itu.edu.tr

© Springer International Publishing Switzerland 2018
A. Ustundag and E. Cevikcan, *Industry 4.0: Managing The Digital Transformation*,
Springer Series in Advanced Manufacturing, https://doi.org/10.1007/978-3-319-57870-5_11

predictability of the outcome and performance of the manufacturing process in real-time, and does not help in autonomously managing and optimizing the cost and time of this process. Moreover, the robots are not able to monitor itself for health issues as part of their self-maintenance ability, and are not able to adapt to a new production process of a new product with different properties because they are designed and built based on the dedicated product, e.g. in automotive industry.

The traditional industrial robots are placed in a designated space and programmed to repeatedly and continuously perform predefined/embedded the same sequence of actions for years and years. Therefore, they are designed, built and equipped for a given sequence of actions, which makes it difficult to reconfigure an industrial robot for a new production line. However, there are also problems of using robots in industry such as the lack of people having expertise and skills to exploit a robot, the hardness of reconfiguring a robot to adapt a new production process, morale on human workers, disabilities of collaborative work on the same space, etc. In addition, the cost of robots is still high even if it continues to decrease.

The term, Industry 4.0 representing the new revolution on industry is based on smart aims to embed the data science into industry in order to generate smart factories for improvement on production/manufacturing. Moreover, many manufacturers think that the production process requires much more collaborative work between robots and humans where robots are intelligent industrial work assistants to humans. To provide such collaborative working, the safety problems for humans are to be solved (Bicchi et al. 2008), thus it is another reason for smart factories to guarantee the safety of humans by controlling the behavior of the robots since the risk of injuring humans may be possible due to collusions. Industrial robots in Industry 4.0 revolution are designed more efficiently and collaboratively with humans and with other robots over networking allowing them to be self-aware and self-adaptable on new products and manufacturing processes (Lee et al. 2015). Thus, the future of industry due to the recent technologies utilizing Internet of Things (IoT) such as for controlling and remotely monitoring on industrial robot (Brizzi et al. 2013), Cloud Computing (Xu 2012; Liu et al. 2014), processing Big Data and advanced information analytics will provide smart factories with many such robots. In addition, the robots are able to autonomously detect degradation on product performance, and apply optimization to solve it. Currently, the lack of predictive information analytic tools is a common problem to process Big Data for industrial issues in order to improve transparency and production quality. Therefore, a framework integrating industrial wireless networks, cloud, and IoT with smart artifacts such as robotic machines is utterly important to implement flexible and reconfigurable smart factory (Wang et al. 2015a, b). Also, process models incorporating novel robots (Erol et al. 2016) are presented to guide companies requiring smart solutions, in their Industry 4.0 vision, and applied on different real-world projects in order to show the need for guided support.

Industry 4.0 represents also an evolution from automated embedded system based manufacturing to Cyber-Physical System (CPS) based manufacturing, which is the recent challenge and requirement for manufacturing industries. CPS's are designed to be equipped with the smart capabilities like sensing, communicating, decision making and actuating to physical world, because such systems making the production decentralized combine the virtual and physical world together using a network in order to enable the robots and humans communicate with each other and share the industrial Big Data collected from the sensors (Lee et al. 2014a, b). The autonomy of the robots is achieved by utilizing the novel components of the information technologies, since these enable the robots sensing and monitoring the production processes, working environment, and even the robots themselves; required for self-predicting, failure recovery, making goal-directed decisions, modelling and controlling the process, self-assessment, etc. One of the benefits of these systems is to make working with robots easier and safer for humans. Therefore, the developments of the sensor technology and networking are essential to provide such capability. Currently, several technical challenges for CPS's such as dealing with uncertainty and dynamics in the real-world, measuring the performance of the systems, lack of full competence on human-machine interaction, etc., most of which are also scientific challenges, are tackled and significant improvements are achieved.

11.2 Recent Technological Components of Robots

The robotic technology, and its development highly depends not only on the cost of materials, but also on the advances of technological components for building a robot making it cheaper, having sensors with higher quality, faster and cheaper processors, the dependence on the open-source robotic software and applications, consuming less energy and being connected everywhere. Moreover, in robotics, there are many scientific challenges such as processing Big Data, dealing with uncertainty, perception in real environment, cognitive decision making in real time, etc. Thus, Industry 4.0 is based on these advances and also the scientific works in the academia to overcome the problems of slow and inefficient decision-making process of autonomous robots, difficulty of using the robots, adopting the robots into manufacturing process, etc. The proposed solutions both from academia and industry comprise the novel hardware and software components of the new age robots.

11.2.1 Advanced Sensor Technologies

For industry 4.0 factories, the advances in the sensor technology has an important role, since these technologies are used in data processing, sharing and collecting. Moreover, networking which remains in the heart of the IoT and Cloud Computing to stream the data depends on the enhancements of wireless sensor technology. In the last decade, many sensors for visual perception, auditory perception, force sensing, obstacle detection, distance sensing, etc., have been developed to be used with robots for industrial tasks by supporting robot's perception such as used in automatic picking, safety handling, part detection, etc.

The advances in the camera technology are important for the performance of image processing, which is an essential ability of the robots for obtaining decisive information about the industrial tasks. Several vision techniques, such as photogrammetry, stereo vision, structured light, time of flight and laser triangulation, utilized for inspection and quality control processes in industrial environments to provide robot guidance are evaluated and compared in terms of accuracy, range and weight of the sensors, safety, processing time and environmental influences (Pérez et al. 2016). This study also investigates vision sensors to be used in the vision system, since the machine vision is an essential ability for many industrial tasks such as obstacles avoidance in the working space and tracking the human workers collaboratively working with the machines. The technology rapidly improves making the cameras smaller, more affordable with high performance, and creates and enhances new aspects for image acquisition such as 3-dimensional (3D) vision, hyperspectral imaging taking a number of images from the same scene in different wavelength ranges, and then combining them to provide depth information. There are various sensors producing Red, Green, and Blue (RGB) depth information and 3D point clouds, covering an infrared projector and a camera such as Microsoft Kinect^{TM}, Asus Xtion^{TM}, etc. Moreover, the sensors including multiple laser sensors like Velodyne^{TM} are used for 3D object detection problems. Such 3D vision endows the robots the capability to achieve aerospace manufacturing tasks such as dispensing designed sealants which are still handled by human operators (Maiolino et al. 2017), by processing the RGB-D data from Asus Xtion^{TM} to use the depth of each pixel with its color information. Also, by using a depth sensor for a marker-less solution for human intention recognition, a multi-modal control approach is developed for human robot cooperation and intuitive communication in order to collaboratively deal with a tightening task (Cherubini et al. 2013).

In addition, to provide safe collaboration between industrial robots and the operators during production, there are many works like in (Bolmsjö 2015) where wearable devices and scanners are utilized. In this study, Epson Moverio BT-200^{TM} glasses is used to obtain information from the environment, the production facility and the robot over Wi-Fi, while allowing the human working with both hands. A safety scanner S300^{TM} from Sick is used to estimate the proximity to a human in order to adapt the robot's speed to the one of the human. The force sensors are also utilized in the industry to obtain tactile feedback while removing and picking material, fitting product parts or assembling complex parts and testing the products in order to detect the collisions (Khalid et al. 2016a, b) and to provide hand guidance for safety movements (Bolmsjö et al. 2016). Moreover, the feedback allows the humans manually controlling the robots. In this study (Rozo et al. 2013), for an assembling task, a KUKA lightweight 7-DoF robot (LWR) equipped with a six-axis force-torque sensor (ATI Mini45^{TM}) is deployed in order to measure the interaction forces generated, while it is holding and moving a wooden table and the human is screwing the legs. In smart factories, operators will be equipped with smart watches, e.g. utilized for lean manufacturing (Kolberg et al. 2015), to receive error messages and error locations during the production.

11.2.2 Artificial Intelligence

Industrial robots in smart factories are to be able to focus on also monitoring, understanding and optimizing the production process, reconfiguring new products, diagnosing and recovering faults. Thus, the robots are designed and developed to have self-awareness, self-maintenance and self-predictiveness abilities. Such abilities are essential capability in a smart factory to improve the production by e.g. ensuring the robot's own health monitoring like detecting a fault existing on torque, or by estimating configuration parameters during production. However, such abilities have been not fully implemented in the real production processes yet. The recent technology makes robots being able to listen to the commands and to perform them without assessing the viability, necessity or rationality of it, and without querying the effects of the process, so the intended self-assessment ability is far from being realized, and it is valid for the other ones. Utilizing the facilities of Cloud Computing and analytic of Big Data, a predictive manufacturing system enables machines and systems with self-aware capabilities, and CPSs are used for the future industry for the improvement of efficiency and productivity (Lee et al. 2013). Intelligent capability based on vision sensing endowed to an industrial robot in order to accurately regulate its own position under uncertainties in the external environment such as calibration errors, misalignment of work pieces, etc., and in the industrial robot such as a mismatch of dynamics, mechanical defects such as backlash, is achieved (Huang et al. 2016).

The autonomous industrial robots have an important role in the smart factories not only to assist the human workers, but also thanking to CPS's, to monitor the everything about a production process due to have the abilities of prediction, decision making, reconfiguring the manufacturing process, optimization of the production line and minimization of energy consumption for improving flexibility and efficiency of the production, economical gains, failure recovery, its own health issues, etc. For the Industry 4.0 factories, several studies are proposed about the intelligent frameworks for negotiation of multi-agent (Wang et al. 2016), collaborative works with human-robot interaction, and predicting the energy consumption for optimization of it. In another study, a modular middleware platform is developed, encompassing robots, a cloud, actuators and sensors using Robot Operating System (ROS) and Artificial Neural Network (ANN) based learning system in order to control Kuka Youbot[TM], which is a factory robot performing preprogrammed actions (Coninck et al. 2016). The intelligent capabilities developed for industrial robots (Zhang et al. 2017) are self-adaptiveness to monitor processes and adapt to the detected disturbances, self-organization to maximize the autonomy and to increase the systems' responsiveness, flexibility, re-configurability, and autonomy.

11.2.3 Internet of Robotic Things

The facilities of IoT for manufacturing enable the communication and interaction among devices, machines and humans in order to obtain, process, analyze, collect and share the knowledge about environment, materials, machines, and processes during production. Ray proposed an architecture of Internet of Robotic Things divided into five layers: (1) the hardware/robotic things layer consisting of various physical things such as robots, sensors, devices, vehicles, etc., to send the information to the network layer, (2) the network layer providing different kinds of networking options such as Wi-Fi, Bluetooth, broadband global area network, etc., (3) the internet layer, which has an important role to provide the whole communication, (4) infrastructure layer providing a framework including the approaches of IoT based robotic cloud, Big Data, middleware and business process, and (5) the application layer (Ray 2017). There are various works based on the pursuit of robotic applications utilizing IoT facilities, such as the case study (Brizzi et al. 2013) for controlling the industrial robot, and monitoring the energy and water consumption using a middleware infrastructure relying on IoT technology in order to integrate industrial sensors and devices over wireless connection. Furthermore, another study (Murar et al. 2015) introduces an approach for controlling remotely a dual-arm industrial robot by using a commercially available IoT technology, ioBridge, to make the monitor and control module, io-2014 which is connected to the robot controller I/O board, communicate with each other.

11.2.4 Cloud Robotics

In smart factories, gathering data along the industrial tasks, analysis of Big Data to extract meaningful information, transferring the data to another robot in different smart factory working on a similar task with the similar sensors are the important challenges of Cloud Computing and networking. Furthermore, an autonomous robot requires the knowledge to perform its operations or for its abilities such as self-assessment, self-configuration, self-maintenance, etc., in order to achieve the tasks and to improve the production quality. However, the knowledge obtained from its sensors and consequently experiences required for these abilities is not sufficient in the smart factory concept. Therefore, the intelligence of the robot is supported by the technologies of Big Data and Cloud Computing (Kehoe et al. 2015) which enable the robots accessing the knowledge from databases, publications, models, benchmarks, simulation tools, open competitions for designs and systems, and open-source software. A cloud based system (Vick et al. 2015) is proposed to provide flexible motion planning and control of an industrial robot by allowing the robot to rapidly reconfigure its control modules, and it is evaluated using a robot manipulator in terms of control performance, availability and scalability. In this work, it has been shown that the advantages of a cloud-based system are to transfer the algorithms requiring high computation time or memory usage to the cloud, to use decentralized system with Internet access, and to be able to use different path planning services at the same time.

In addition, a cloud robotics based application is developed to connect physical resources such as robots, devices and sensors in a cloud over a local server, and it is evaluated by two case studies (Wang et al. 2017): In the first one, by applying collision detection using 3D point clouds of the robot and human captured by depth cameras for online control of the robot, safe and protected environment for human operators are provided. In the second study, the energy consumption of an industrial robot's movements is minimized by estimating the most energy-efficient joint configuration. For these studies, a local collision avoidance server in the cloud is utilized allowing the communication between the robot, the cameras, and a local server to fine-tune the most optimal robot parameters, such as the estimate of the efficient inertial tensors of the robot's joints.

A customization manufacturing system including three layers; (1) a manufacturing device layer to allow various devices such as an industrial robot to obtain decisions about the production in the cloud, (2) a cloud service system layer, where the concept of a CPS is qualified, and (3) a mobile system layer to provide a terminal in order to access the cloud is designed (Wan et al. 2016). A platform covering a number of physical devices connected to a cloud via wireless network, which are a conveyor belt, multiple industrial robots, and various sensors is implemented. The cloud provides the required data inquired from the customers about the production, to be processed and to make decisions related to the customization manufacturing.

11.2.5 Cognitive Architecture for Cyber-Physical Robotics

Cyber-physical systems are the embedded systems equipped with physical systems and environments through the combinations of computational modules. The application areas of the systems cover not only smart machines and factories, but also smart transportation, smart buildings, smart cities, smart medical technologies, etc., thus, the integration of such systems with the technologies like IoT, Big Data analytic, Cloud Computing and wireless sensor networking is the core of the Industry 4.0 revolution. The main components of the system infra-structure are actuators to interact with these world, the sensors for sensing the physical environment, and information processing, therefore, the advances on these related technologies will determine the development of the CPS's.

A 5C architecture including 5 levels, which are (1) Connection as the level to manage data acquisition systems such as sensors, data sources and transferring protocols, (2) Conversion, where the data is processed and transformed into valuable and usable knowledge, (3) Cyber as central information hub in order to establish cyber space using the information acquired from every source, (4) Cognition for optimization of the decisions and (5) Configuration as supervisory control to endow machine self-configure and self-adaptive, is designed to integrate cyber-physical systems with smart machines to be used the manufacturing industry (Lee et al. 2015). Thus, such systems are the core of a smart factory due to the incorporation of IoT and Big Data with physical industrial world.

There are many CPS studies in industrial tasks like in where an integrated cyber-physical system is proposed (Wang et al. 2014) by building cloud-based services of monitoring, planning of production processes, machining and assembly in decentralized environment, for remotely accessing and controlling an equipment used in production such as Computer Numerical Control (CNC) machines and robots. The CPS software plays the most important role on the development of this kind of system; related issues like the analysis, design, development, verification and validation, and quality assurance of CPS software need to be taken into account (Al–Jaroodi et al. 2016). In addition, A CPS in human robot collaboration also called Collaborative Robotic CPS (CRCPS) covers three main integrated entities; (1) Human Component (HC) connected through different adaptor technologies, e.g. accurate human position tracking technology, (2) the Physical Component (PC) and (3) the Computational Component (CC). A CRCPS is developed considering safety and protection measures in order to increase productivity (Khalid et al. 2016a, b). Such a system is able to utilize a variety of sensors and actuators, and is intended to provide the interaction between HC, CC and PC, for example a vision system for detection, tracking and gesture recognition of human workers.

11.3   Industrial Robotic Applications

Industrial robots with various intelligent and sensory capabilities are utilized in the manufacturing processes. In an Industry 4.0 factory, the robots endowed with the advanced capabilities owing to the information, networking and sensor technologies are able to collaboratively work with human workers and cooperatively with the other robots in an assembly line. The collaborative and cooperative working applications of the robots, the maintenance practices and assembly line applications using the robots shape the factories of the future.

11.3.1   Manufacturing

Conventional manufacturing industry was transformed into manufacturing by intelligent systems in order to improve production performance and economical benefit in many advanced countries (Lee et al. 2014a, b). The machines like industrial robots instead of humans were already utilized a few decades ago in order to enable faster and more accurate production than using human workers. However, in recent industries, manufacturing is not sufficient due to disabilities of robots and the hardness of implementations of current technological development.

Due to the benefits of the information and sensory technologies, the enhancement of the interaction between the robotic machines and the human workers as well as the robots themselves will improve the manufacturing quality.

11.3.1.1 Human-Robot Collaborative Manufacturing

In the recent industrial manufacturing, the requirement of industrial robots and human workers for collaborative works over a communication network (Clint 2010) is growing, and such communication is achieved due to the advances of the aspects of Industry 4.0. Moreover, it is also believed that integrating the industrial robots into human working spaces makes the production process more economical, and enables many collaborative applications in the factories (Bahrin et al. 2016). However, the robots and humans are not much capable of smoothly working together in the same working space because of safety issues. Even if the robots are designed to be cautious around human workers, it is aimed to achieve a more flexible and agile moving capability for a robot, which is highly required to provide more efficiency in working collaboratively with the humans. The collaborative robots are still not maturely developed to work for complex tasks in the same space with humans in assembly lines and distribution centers. Lenz et al. designed a concept for a smart factory, allowing the joint-actions of humans and industrial robots, where robots perceive their environment with multiple sensor modalities is designed to collaboratively assemble capital goods (Lenz et al. 2008).

In addition, there are many studies aiming for the collaboration of humans and robots while considering the safety of the human workers in which a human-robot interaction system is proposed to guarantee the safety by tracking and estimating the proximity of the workers to the robots, and by activating the strategies considering the proximity (Corrales et al. 2012). Khalid et al. investigated in their comprehensive study the collaboration mechanisms based on CPS, in which a variety of sensors and actuators are used to enable the interaction among the HC, CC and PC entities, considering several indices such as estimating the safety distance based on different sensors for human position monitoring, cost, risk, collaboration category, performance level, etc., (Khalid et al. 2016a, b). The sensors are selected considering the safety concepts, such as force monitoring, speed and distance monitoring, and the complete isolation. Also, in this study, the collaboration is formalized taking into account the number of sensors, the data rate from the sensors to calculate the key performance indicators, and considering the speed of the human workers in the same working space approaching to collision with the robot, and finally the robot's follow-up time to stop completely by calculating the safe distance in order to provide the safety of human workers.

To process and extract meaningful information about a specific task from Big Data generated by a collaborative community including many robots and humans makes the robots able to cope with and explain uncertainties, and to make better decisions. This kind of industrial Big Data may not only consist of knowledge obtained from this community, but also knowledge transferred from other robots in different factories over internet. One of the most important challenge in collaborative working of humans and robots is how to transfer information because the robots should be able to know in advance where the humans are and what they do in the common working space (Augustsson et al. 2014). However, there is no such smart analytic and predictive tools to gather, to process and to transfer Big Data (Wang et al. 2015a, b). In the future, to provide advanced collaboration between human workers and industrial robots, CPS will be developed for dynamic task planning, active collision avoidance, and controlling the robots by speech, gestures and signs (Wang et al. 2015a, b).

11.3.1.2 Cooperating Robots in Manufacturing

In addition to collaborative working of the machines and humans, in the smart factories, the robots need to be capable of working together with other robots connected on a collaborative community, and to share the information gathered from their own work over this community. Such sharing depends on the enhancement on the wireless networking, and is to improve the robotic abilities of the learning, prediction, flexibility and the decision-making on the production process. For the Industry 4.0 factories, several studies are proposed about the intelligent frameworks integrating a smart factory framework with multiple autonomous agents with Big Data based system (Wang et al. 2016) consisting of physical resource layer, industrial network layer, cloud layer, and supervisory control terminal layer, for distributed self-decision making and intelligent negotiation mechanisms of the agents to implement self-organized manufacturing system in order to prevent deadlock. Also, in another study (Wagner et al. 2016), the cooperative working of two industrial robotic arms is shown, where one is used for the sensory task, which is to scan the work pieces like pins and to inform the other one handling the task, by creating point clouds and modifying it during the motion.

Wang et al. has shown a smart factory platform experiment with robots used for industrial tasks, having wireless connection and a private cloud (Wang et al. 2015a, b). A number of serial robots for loading tasks and Cartesian robots for machining and testing tasks, and a railway like conveying system were used. A serial robot serves two Cartesian ones equipped with manipulators to achieve different tasks. The serial ones put the products on Cartesian ones, which are brought from the conveying system, and vice versa. These are also able to autonomously control their position and velocity, and able to avoid collisions. Moreover, the robots are equipped with smart controllers and devices required to make them the components of the smart factory. The software provides the capabilities for cloud computing and Big Data analytic, and virtualizes the network of servers as a supercomputer of the platform, where the robots constitute the clients. By using this network architecture, the smart factory is constructed where the robots can communicate with each other, the massive data obtained from tasks and experiences can be collected and transferred to cloud, and over the cloud, the big data can be transferred to other robots. Also, a ROS based framework for cyber-physical production systems is proposed in order to develop and improve the coordination strategies for cooperating robots considering network delays, localization inaccuracies, and availability of embedded computational power (Böckenkamp et. al. 2016). Such a framework makes the robots more adaptive and cooperative since it combines collision detection and avoidance using local decisions based on observations from the environment.

11.3.2 Maintenance

In factories, maintenance of a production process is one of the most essential planning items to diagnose the health problems of the machines and to reduce the downtime of the process due to these problems. Humans are still responsible for this task in the recent technology, but several systems are proposed to enable the machines having self-maintenance ability. For instance, Prognostics and Health management (PHM) (Vichare et al. 2008) is a discipline for machines and robots to assess the health of systems in order to diagnose anomalies using sensors, and in order to predict the performance over the life of the machines. The development of IoT provides opportunities for the discipline PHM to be used efficiently in manufacturing (Kwon et al. 2016), such as the speed of decision-making, improved reliability and accountability over a cloud, enhanced liability and workforce competency.

To track and to assess its own health and overall production performance degradation, self-awareness and self-maintenance are required abilities for the robots. By processing the information from Big Data, industrial robots are able to manage their health and maintenance. In a recent study (Pedersen et al. 2016), it is shown how robot skills for manufacturing and new tasks can be derived by transferring information from factory worker's knowledge in order to develop self-asserting ability. For the robot maintenance and fault diagnosis, the vibration on an industrial robot is measured using accelerometers and evaluated by applying Fast Fourier Transform for the analysis of the auditory spectrum of vibration (Vagaš et al. 2014). Also, the tooth failures of the gears on an industrial robot such as scuffing, cracking, macro- and micro-pitting, wear, bending fatigue, and fracture are diagnosed using Discrete Wavelet Transform and Artificial Neural Networks (Jaber et al. 2016).

11.3.3 Assembly

Conventional modern assembly lines are highly automated, but not dynamically adaptive to new production requirements. However, it is notoriously difficult to make the robots easily reconfigured and reprogrammed to the changes in production lines (Lee et al. 2014a, b). Manufacturers feel the need of an assembling process, in which the humans and robots work together, and the optimization of the production process and reduction of the downtime are achieved autonomously by monitoring the assembly tasks and making decisions. Such intelligent systems are possible in the Industry 4.0 factories.

Adapting new production paradigms and reconfiguring automation, tasks in assembly line during production is another requirement for the industrial robots in a smart factory. Within the framework of Industry 4.0, Pfeiffer explains how assembly works will change by focusing on the non-routine and neglected works in core assembly tasks, and what the roles of humans and robots will be in an assembly task considering their interactive capabilities to ensure high performance, quality, and a smooth material flow (Pfeiffer 2016). Also, it is needed to utilize the Industry

4.0 technologies, such as IoT, in the development of robots for all kind tasks of an assembly line such as the assembly conveyor tasks (Berger et al. 2015) in which the robot is developed to be able to achieve manufacturing task for synchronizing the industrial mobile robot with the moving object used in this process. In this study, there are two systems to achieve the synchronization, to control the mobile platform, to localize the object, on which three LEDs need to be mounted, as well as the mobile platform by relying on an optical sensor to estimate the position of each LED.

A number of distributed and collaborative CPSs are designed and utilized to create a feature-based manufacturing process as a cyber-physical robot application (Adamsona et al. 2017). This process is used for equipment control of robots and matching resources and tasks in order to achieve assembly tasks by combining assembly features of products and event-driven function blocks. The features consist of the coordination and control of a set of robot motions and actions such as signal processing, program logic, decision-making, and communication. Thanks to the CPSs, manipulations of the components of a product are related and the assembly method is applied to the product considering the assembly features in this assembly task scenario.

11.4   Conclusion

In this chapter, it is summarized how the facilities of the Industry 4.0 technologies covering sensor, networking and information technologies are incorporated to the industrial robots for manufacturing tasks, how these facilities affect the production and what the requirements of the manufacturers on the technologies are. Experimental works using the industrial robotic systems for the smart factories in assembly tasks, networking, collaborative and cooperative works with humans are surveyed. As a result of the integration of advanced sensors with Artificial Intelligence, IoT and Cloud Computing technologies, which will be utilized more commonly in the factories, the industrial robots having intelligent decision-making, prediction, and maintenance abilities with advanced autonomous behaviors will contribute more to the production by helping to and working alongside with the human workers.

References

Adamson G, Wanga L, Moore P (2017) Feature-based control and information framework for adaptive and distributed manufacturing in cyber physical systems. J Manufact Syst 43:305–315
Al-Jaroodi J, Mohamed N, Jawhar I, Sanja LM (2016) Software engineering issues for cyber-physical systems. In IEEE international conference on smart computing (SMARTCOMP), pp 17–23
Augustsson S, Olsson J, Christiernin LG, Bolmsjö G (2014) How to transfer information between collaborating human operators and industrial robots in an assembly. NordiCHI 2014:286–294
Bahrin MAK, Othman MF, Azli NHN, Talib MF (2016) Industry 40: a review on industrial automation and robotic. Jurnal Teknologi 137–143

Berger U, Le D, Zou W, Lehmann C, Stdter J, Ampatzopoulos A (2015) Development of a mobile robot system for assembly task on continuous conveyor. In International conference on innovative technologies, pp 331–334

Bicchi A, Peshkin MA, Colgate JE (2008) Safety for physical human robot interaction. In: Springer Handbook of Robotics Heidelberg, Germany, Springer, pp 1335–1348

Bolmsjö G (2015) Supporting tools for operator in robot collaborative mode. In 6th international conference on applied human factors and ergonomics (AHFE 2015) and the affiliated conferences, AHFE 2015, pp 409–416

Bolmsjö G, Bennulf M, Zhang X (2016) Safety system for industrial robots to support collaboration. Adv Ergonom Manufact: Manag Enter Fut 490:253–265

BöckenkampFrank A, Stenzel W, Lünsch D (2016) Towards autonomously navigating and cooperating vehicles in cyber-physical production systems. In Machine learning for cyber physical systems, pp 111–121

Brizzi P, Conzon D, Khaleel H, Tomasi R, Pramudianto F, Knechtel M, Cultrona P (2013) Bringing the internet of things along the manufacturing line: a case study in controlling industrial robot and monitoring energy consumption remotely. In Emerging technologies & factory automation (ETFA) pp 1–8

Cherubini A, Passama R, Meline A, Crosnier A, Fraisse P, (2013) Multimodal control for human-robot cooperation. In 2013 IEEE/RSJ international conference on intelligent robots and systems (IROS), pp 2202–2207

Clint H, (2010) Human-robot interaction and future industrial robotics applications. In IEEE/RSJ international conference on intelligent robots and systems, pp 4749–4754

Coninck ED, Bohez S, Leroux S, Verbelen T, Vankeirsbilck B, Dhoedt B, Simoens P, (2016) Middleware platform for distributed applications incorporating robots. In Sensors and the cloud, 5th IEEE international conference on cloud networking

Corrales JA, García Gómez GJ, Torres F, Perdereau V (2012) Cooperative tasks between humans and robots in industrial environments. Int J Adv Rob Syst 9(94):1–10

Erol S, Schumacher A, Sihn W (2016) Strategic guidance towards industry 40—a three-stage process model. In International conference on competitive manufacturing (COMA 2016), pp 495–500

Huang S, Bergström N, Yamakawa Y, Senoo T, Ishikawa M (2016) Applying high-speed vision sensing to an industrial robot for high-performance position regulation under uncertainties. Sensors 16(8):1195

Jaber AA, Bicker R (2016) Fault diagnosis of industrial robot bearings based on discrete wavelet transform and artificial neural network. In-Non-Dest Test Cond Monit 7:179–186

Kehoe B, Patil S, Abbeel P, Goldberg K (2015) A survey of research on cloud robotics and automation. IEEE Trans Autom Sci Eng 12(2):398–409

Khalid A, Kirisci P, Ghrairi Z, Pannek J, Thoben KD (2016) Safety requirements in collaborative human robot cyber physical system. In 5th international conference on dynamics in logistics (LDIC 2016), pp 39–48

Khalid A, Kirisci P, Ghrairi Z, Thoben KD, Pannek J (2016) A methodology to develop collaborative robotic cyber physical systems for production environments. Log Res 9. doi: 10.1007/s12159-016-0151-x

Kolberg D, Zhlke D (2015) Lean automation enabled by industry 40 technologies. IFAC-PapersOnLine 48(3):1870–1875

Kwon D, Hodkiewicz MR, Fan J, Shibutani T, Pecht MG (2016) IoT-based prognostics and systems health management for industrial applications. IEEE Access 4:3659–3670

Lee J, Bagheri B, Kao HA (2015) A cyber-physical systems architecture for industry 40-based manufacturing systems. Manuf Lett 3:18–23

Lee J, Kao HA, Yang S (2014a) Service innovation and smart analytics for industry 4.0 and big data environment. Procedia CIRP 16:3–8

Lee J, Bagheri B, Kao HA (2014) Recent advances and trends of cyber-physical systems and big data analytics in industrial informatics. In Proceedings of International Conference on Industrial Informatics (INDIN), pp 217–229

Lee J, Lapira E, Bagheri B, Kao HA (2013) Recent advances and trends in predictive manufacturing systems in big data environment. Manuf Lett 2013:38–41

Lenz C, Nair S, Rickert M, Knoll A, Gast J, Bannat A, Wallhoff F, (2008) Joint-action for humans and industrial robots for assembly tasks. In Proceedings of the 17th IEEE international symposium on robot and human interactive communication, pp 130–135

Liu Q, Wan J, Zhou K (2014) Cloud manufacturing service system for industrial-cluster-oriented application. J Int Technol 28(1):373–380

Maiolino P, Woolley R, Branson D, Benardos P, Popov A, Ratchev S (2017) Flexible robot sealant dispensing cell using RGB-D sensor and off-line programming. Robot Comput Int Manuf 48:188–195

Murar M, Brad S (2015) Monitoring and control of dual-arm industrial robot tasks using IoT application and services. Appl Mech Mater 762:255–260

Pedersen MR, Nalpantidis L, Andersen RS, Schou C, Bøgh S, Krüger V, Madsen O (2016) Robot skills for manufacturing: from concept to industrial deployment. Robot Comput-Int Manuf 37:282–291

Pérez L, Rodríguez Í, Rodríguez N, Usamentiaga R, García DF (2016) Robot guidance using machine vision techniques in industrial environments: a comparative review. Sensors 16(3):335

Pfeiffer S (2016) Robots, industry 40 and humans, or why assembly work is more than routine work. Societies 6(2):16

Ray PP (2017) Internet of Robotic Things: Concept, Technologies, and Challenges. IEEE Access 4:9489–9500

Rozo LD, Calinon S, Caldwell D, Jimenez P, Torras C (2013) Learning collaborative impedance-based robot behaviors. In Proceedings of the twenty-seventh AAAI conference on artificial intelligence, pp 1422–1428

Vagaš M, Semjon J, Baláž V, Varga J (2014) Methodology for the vibration measurement and evaluation on the industrial robot KUKA. In Proceedings of the RAAD 2014 23rd international conference on robotics in Alpe-Adria-Danube region

Vichare NM, Pecht MG (2008) Prognostics and health management of electronics. Wiley, Hoboken

Vick A, Vonasek V, Robert P, Kruger J (2015) Robot control as a service—towards cloud-based motion planning and control for industrial robots. In Proceedings of 10th ieee international workshop on robot motion and control (RoMoCo), pp 33–39

Wagner M, Hess P, Reitelshoefer S, Franke J (2016) 3D scanning of workpieces with cooperative industrial robot arms. In Proceedings of ISR 2016: 47th international symposium on robotics

Wan J, Yi M, Li D (2016) Mobile services for customization manufacturing systems: an example of industry 40. IEEE Access 4:8977–8986

Wang L, Gao R, Ragai I (2014) An integrated cyber-physical system for cloud manufacturing. In Proceedings of the ASME 2014 international manufacturing science and engineering conference, vol 1. pp 135–144

Wang L, Törngren M, Onori M (2015) Current status and advancement of cyber-physical systems in manufacturing. J Manuf Syst 37(2):517–527

Wang S, Wan J, Li D, Zhang C (2015b) Implementing smart factory of industrie 40: An outlook. Int J Distrib Sens Netw 2016:7–17

Wang S, Wan J, Zhang D, Li D, Zhang C (2016) Towards smart factory for industry 40: A self-organized multi-agent system with big data based feedback and coordination. Comput Net 101:158–168

Wang XV, Wang L, Mohammed A, Givehchi M (2017) Ubiquitous manufacturing system based on cloud: a robotics application. Robot Comput-Int Manuf 45:116–125

Xu X (2012) From cloud computing to cloud manufacturing. Robot Comput-Int Manuf 28(1): 75–86

Zhang Y, Qian C, Lv J, Liu Y (2017) Agent and cyber-physical system based self-organizing and self-adaptive intelligent shopfloor. IEEE Trans Indust Inf 13(2):737–744

Chapter 12
The Role of Augmented Reality in the Age of Industry 4.0

Mustafa Esengün and Gökhan İnce

Abstract Augmented Reality (AR) has increased its popularity in both industry and academia since its introduction two decades ago. The AR has carried the way of accessing and manipulating the information to another level by enhancing perception of the real world with virtual information. In this chapter, the basic technical components of AR are introduced as well as its practical uses in the industry especially in manufacturing, maintenance, assembly, training and collaborative operations. After describing the underlying hardware and software systems establishing AR, recent applications of AR in the industry are reviewed.

12.1 Introduction

Human beings perceive their environment as much as the capabilities of their five senses allow. Humans cannot see what is not in their environment or cannot touch an object which does not exist physically. Therefore, the interaction with the outer world and access to the relevant information about any given task is rather limited.

For example, a junior engineer who fixes an automobile engine cannot troubleshoot the malfunction that s/he does not have knowledge about without conducting to an expert or a manual. In short, we need other means when we need extra information about the items that we interact in our daily life. These means could be internet, a paper based manual, a colleague, etc. The way these means are accessed also affects the quality of the interaction or experience with the objects. For an enhanced interaction, humans need to get the nonvisible information related to the interacted object as quickly as possible and within an understandable form. Augmented Reality (AR) technology has been considered as an innovative way of

M. Esengün (✉) · G. İnce
Faculty of Computer and Informatics Engineering, Department of Computer Engineering,
Istanbul Technical University, 34469 Maslak, Istanbul, Turkey
e-mail: esengun@itu.edu.tr

G. İnce
e-mail: gokhan.ince@itu.edu.tr

© Springer International Publishing Switzerland 2018
A. Ustundag and E. Cevikcan, *Industry 4.0: Managing The Digital Transformation*,
Springer Series in Advanced Manufacturing, https://doi.org/10.1007/978-3-319-57870-5_12

interaction for this purpose. Thanks to the AR, humans can explore more than that of their five senses perceive. The main aim of AR is to enhance a humans perception about the environment by overlaying additional computer generated visual information onto the vision of the user through specific devices, such as, camera of the smartphones, head-mounted displays (HMD), projection devices, etc. The computer generated visual information can be images, videos, 3D models, texts, sounds, speech instructions etc. (Boulanger 2004). By superimposing this type of augmentation onto the real vision of the user, the user is able to get the hidden information about the interacted objects or the environment, which makes the AR an efficient technology for several areas, such as, gaming, sports, advertising, shopping, education, military services, medical surgeries as well as industrial purposes, etc.

Three key properties of AR are defined as presenting virtual and real objects together in a real environment, allowing interaction with virtual and real objects in real time, and registering (aligning) virtual objects with real objects (Azuma et al. 2001). The additional information brought by the augmentation could assist a user when performing real world tasks. With these properties, AR can significantly contribute to the needs of industry. Moreover, thanks to the advances in hardware and software systems, AR has been attracting more attention from industry than before.

12.2   AR Hardware and Software Technology

In general, an AR system contains four hardware components; (1) a computer, (2) a display device, (3) a tracking device and (4) an input device (Wiedenmaier et al. 2003). The computer is responsible for not only modelling augmentations and controlling all the connected devices but also adjusting the position of augmentations in the real scene with respect to the position of the user by using the information gathered from the tracking device. A display device is necessary to display the augmentations on top of the user's real vision. The choice of the display device depends on the type of interaction. The most widely used technologies are a see-through Head-Mounted Display (HMD), which the user wears on his head, a Hand-Held Display (HHD), such as a tablet or a smartphone, or Spatial Displays (SD), which is designed using several projectors. The tracking device is responsible for tracking exact position and orientation of the user, and then registers the augmentations properly to their desired positions. Using the tracking device, the computer provides assistance to the user by processing the user's movements, gestures and actions while the user interacts with the objects in the real world (Rose et al. 1995). The input device is used for enabling the user to interact with the system. Some examples of the input devices are microphone, touchpads, wireless devices, mouse and haptic devices (Dini and Mura 2015). The choice of each component depends on the application scenario. If the application is designed for the technicians working in a factory, it is logical to choose wireless see-through

HMDs, since in most of the tasks they need both their hands to complete the job; so HHDs are not efficient in this case. In addition, see-through HMDs provide mobility and are considered more suitable for outdoor applications (Boulanger 2004). However, currently available HMD devices are not much comfortable because of their weights, dimensions and weak resolution. It has been reported numerous times that especially after some time of usage they may cause headache and dizziness (Nee and Ong 2013; Regenbrecht et al. 2005). It is also presented that maintenance operators experienced difficulty because of the bulky HMDs being not suitable to their working environment (Didier et al. 2005). On the other hand, the HHDs are quite powerful in terms of having high resolution camera and various sensors, but they suffer from limited processing power. Therefore, in most of the cases researchers developed a client-server architecture to improve the performance of HHD (Nee and Ong 2013).

Superimposing augmentations onto the user's scene is done in several ways, which also affect the choice of devices. One way is that the augmentations are directly projected to the field of view of the user, which is called optical combination and is implemented with an optical see-through HMD. Another technique is called video mixing. In this method, user's scene is captured by a camera and processed with a computer and after augmentations inserted on the processed scene, the result is displayed on a monitor, on which the user observes the real scene indirectly. The last method is image projection, in which the augmentations are directly projected onto the physical objects (Dini and Mura 2015). For additional information about these methods, the reader is advised to refer to the early work of Azuma (1997) which presents a detailed comparison of optical and video technologies.

Tracking and registration are apparently the most important challenges in AR applications. Only with an accurate tracking and registration, augmentations are aligned correctly. The tracking and registration algorithms are categorized into three classes such as, (1) marker-based algorithms, (2) natural feature-based (or markerless) algorithms and (3) model-based algorithms (Nee and Ong 2013). In marker-based algorithms, 2D markers having unique patterns or shapes are placed on the real objects, where the augmentations are to be superimposed. An augmentation is assigned for each marker in the workplace programmatically and once the camera of the device recognizes the markers, their assigned augmentations are displayed on the markers. In some cases, using markers is not efficient and hence, natural feature-based preprocessing algorithms widely used in computer vision, such as, Speeded Up Robust Features (SURF), Scale Invariant Feature Transform (SIFT), Binary Robust Independent Elementary Features (BRIEF) are preferred. Lastly, Model-based algorithms compare extracted features with a predefined list of models. Today, there are a lot of Software Development Kits (SDK) especially developed for making AR application development easier. Metaio, Vuforia, Wikitude, ARToolKit, and Hololens are the most popular SDKs and provide functionalities and detailed documentations, which give the opportunity for developing AR applications to a developer with low coding skills and experience.

12.3 Industrial Applications of AR

AR provides assistive solutions in various industrial fields making processes easier
to manage, helping in reducing the human error, enabling a new way of educating
people and increasing the collaboration. In this sense, the usage of AR in manufacturing fields is analyzed in this section.

Manufacturing processes focus on satisfying human needs by providing qualified products using raw materials and knowledge. Because of the dynamic and
competitive nature of the today's business world, companies deal with the challenges of managing cost, time, quality and flexibility (Chryssolouris 2006). Since
the companies should release their innovative products in short time durations and
at low cost, their manufacturing processes have to be more responsive and systematic. Moreover, real-time information exchanges between almost all the phases
of product development life cycle, e.g., design, planning, assembly, maintenance,
etc., are also required for decreasing production time and cost as well as massive
customization of the products according to the needs of the customers (Ong et al.
2008). In order to address these time and cost factors as well as collaboration
between phases of the product development life cycle, digital manufacturing has
been considered (Chryssolouris 2006). Incorporating computer support in manufacturing systems not only eases the error handling, but also enhances the
decision-making processes. One of the innovative and effective solutions for
computer supported manufacturing systems is the AR technology. The AR technology has been used in order to improve manufacturing processes by helping in
solving critical problems, or by preventing subsequent re-works and modifications
in the activities, such as, design, planning, etc., before the actual processes are to be
carried out. AR allows users to interact directly with the information related with
the manufacturing processes in real time and in real working environment, which is
highly useful especially in the maintenance, training, assembly, product design,
layout planning and other manufacturing activities (Nee et al. 2012; Ong and Nee
2004; Li et al. 2004).

12.3.1 Maintenance

Maintenance services are one of the most crucial processes of manufacturing.
A maintenance operation involves activities such as, analysis, testing, servicing,
alignment, installation, removal, assembly, repair, or rebuilding of human made
systems (Henderson and Feiner 2009). Since the quantity of the functionalities of
the products has been increasing, the products become more and more complex,
which requires even the most experienced users to consult the manuals, prints, or
computers to retrieve information about safety, maintenance procedures, or component data (Henderson and Feiner 2007; Ke et al. 2005). From the perspective of
technicians, when they face with a problem on site, they usually use paper manuals

or get instructions from other experts by phone calls or on site. However, it is a fact that solving problems by using paper manuals or talking with experts is an inefficient way in terms of cost and time, which may even finally result in customer dissatisfaction (Wang et al. 2014; Nee et al. 2012). Moreover, referencing paper manuals or consulting to an expert decouples the focus of the technicians from the repair area of the product by filtering information related to the problematic part of the equipment and trying to match that information with the real product, which also increase the work load of the technicians (Henderson and Feiner 2007). Another problem of the paper manuals is that they change over time as the products are modified during their lifecycle. Therefore, the procedures and drawings presented in the paper manuals have to be updated, which also cause extra cost and time issues (Oliveira et al. 2014). In order to solve this problem, instructions presented in the manuals are made available virtually in an AR system with the updated information about the products (Oliveira et al. 2013).

AR technology has been widely used to support maintenance activities. In one of the earliest works Raczynski and Gussmann (2004) specified maintenance activities and features, where AR could be involved as a support mechanism. For the analysis phase, where the systems detect the components which need maintenance, virtual information could be overlaid on the image of the real situation, displaying steps for inspection. After the inspection step, a service worker needs to be supported with fault diagnostics by providing communication with the expert, so that, the field of view of the worker at the fault component can be transmitted to the remote expert. The remote expert can then support the worker by using text annotations, prepared arrows or other geometrical shapes to show the necessary actions that the worker need to make. Finally, the worker handles the problem by following the augmented instructions that are provided by the expert.

Supporting maintenance tasks with AR applications could keep the users in the task domain without continually switching their focus of attention between the task and a separate manual (Henderson and Feiner 2007). Superimposing the necessary information about the task in real working environment is one of the most distinct features, which makes AR essential for maintenance tasks. The superimposed information could be in the form of text annotations, animations, videos, images or computer-generated 3D models. Therefore, by using AR applications users are provided with more content than with the paper manuals. In such tasks, the user is equipped with a wearable computer unit and gets the necessary information overlaid onto the object being inspected, showing needed actions step-by-step and how to do them. Hence, the amount of errors and task completion times decrease (Ke et al. 2005; Azuma 1997; Regenbrecht et al. 2005; Haritos and Macchiarella 2005). Other than reduced frequency of context switching and time for maintenance tasks, AR also enables real time collaboration. The technician's field of vision can be transferred to an expert over a shared network. The source of the live video is in that case a head camera worn by the technician. The transferred video data enable the remote expert to monitor the problem and to give necessary instructions by sharing augmentations transferred back to the technician over the network. This way of real time collaboration also makes the technicians retain how they handled the problem better than remembering by using separate manuals in case they face with the same problem in the future (Henderson and Feiner 2007). AR based collaborative maintenance allows both the remote expert and the field technician to use some of the spatial cues used in face-to-face collaboration that are normally lost in remote conference systems, which results in higher social presence than other technologies (Lukosch et al. 2015).

There have been various projects focusing on AR-based maintenance applications. Rose et al. (1995) presented an AR application which is used for annotating parts of an automobile engine. In the developed system, the user can point any part of the engine with a pointing device and see the augmented information related to the pointed part by wearing a see-through HMD. The augmentations were simple text and lines that connects the text to the part it describes. The authors stated that by augmenting the textual annotations, the user is able to be informed about the tasks and parts in more detail without separating the focus of the user from the product. Moreover, the dangerously hot and electrified parts of the product can also be highlighted to prevent danger of touching them. The ARMAR project was developed to aid technicians for maintenance tasks (Henderson and Feiner 2007). The system tracks the users head position by using a tracking device and uses a see-through HMD to display the interface, which presents augmentations to the technician. The augmentations inform the user about the problem and necessary steps for the solution as texts or 2D images. One important feature of this project is that since the use of HHDs causes some restrictions especially when the technician needs to use his hands, gesture recognition was provided using a second camera which enhances the user experience and create opportunities for designing user interfaces including virtual buttons on augmented view.

The transfer of the live images of the local technicians view to the remote expert has been handled by implementing networking solutions. Wang et al. (2014) presented such a network environment to achieve efficient collaboration between a technician and an expert located in different places. In the developed framework, if the technician needs assistance with a specific component of the product, s/he pastes markers onto these components and asks maintenance questions by text or oral communication through a network connection. The view of the technician is captured by a camera and the images captured by the camera including marker information are transmitted to the remote expert through the network in real time. On the remote expert site the markers are recognized and the expert adds virtual content, such as, texts, arrows, model of the components in order to create a maintenance scenario. Then, the constructed scene is transmitted to the technicians display as augmentations. Moreover, augmentations play an important role for the users to understand details of especially complex products. Even the simplest form of the augmentations can improve the quality and efficiency of the maintenance processes. A simple desktop PC maintenance application is demonstrated by Ke et al. in 2005. The user is equipped with a see-through HMD, which includes a camera for detecting the components of the desktop PC. If the user needs support for a specific component on the PC, s/he pastes physical markers on these components and the system displays augmentations overlaid onto these markers involving maintenance information. The maintenance process starts with collecting the image of the desktop PC in the reality by using the camera and transmitting the image information to a separate computer, which reconstructs the image by superimposing virtual information including names of the components of the PC on their related positions. Once transmitting the final image to the head-mounted display, the user can finally perform necessary actions by the help of the augmentations.

Tracking technique of AR systems is crucial in ensuring their acceptance in the industry. The use of markers for tracking and identifying parts of a complex product may not be efficient in some cases since markers occlude parts of the workspace and require a free line of sight with the camera, which limits the technicians movement. Therefore, eliminating markers may be considered more natural and user friendly in most of the situations. An example of markerless tracking approach was implemented by the Metaio GmbH for maintenance operations in automotive industry (Platonov et al. 2006). This project features an optical see-through display which communicates wirelessly with an AR application running on a standard notebook. The proposed markerless tracking algorithm is based on 2D point features of the video images fetched from camera mounted on the HMD and a preprocessed computer-aided design (CAD) model of the equipment being maintained. Once the signal of the video camera is transmitted to the notebook, where both tracking and augmentation happen, the complete augmented video stream providing virtualized repair instructions, part labeling and animation for maintenance steps is sent back to the HMD. This system allows users to perform the maintenance tasks without any intervention to the environment.

Besides the tracking techniques of the AR systems, choice of the device used for displaying augmentations is also important for user experience. Since most of the head-mounted displays contain various hardware components such as sensors and cameras, they are bulky devices with high weight and may even have narrow field of vision with low resolution. To deal with this issue, a robotic arm carrying a camera and a pico projector for receiving the users view and project augmentations onto the user's workspace was proposed (Gurevich et al. 2012). The user can direct the robotic arm to the point where s/he specifically needs help and the camera on the robotic arm transmits the live video of the workspace to the remote expert through network. After analyzing the scene of the worker, remote expert can draw texts, pointers or other graphics highlighting the instructions for the worker and send the information back to the projector on the robotic arm to display onto the users workspace.

12.3.2   Assembly

Assembly operations involve the manipulation and joining of separate parts to form a whole (Nevins and Whitney 1980). Especially for complex products, where the number of parts are high, assembly operations can become rather difficult tasks to manage. As in the other maintenance processes, assembly operations also involve several steps, such as, (1) analysis, in which the worker visits the site and gathers all the information related to the task and do planning for the assembly; (2) diagnostics, in which the worker adapts the support for the task to his skills; and (3) therapy, in which the prepared support tools involving instructions for the assembly task are handed to the worker (Raczynski and Gussmann 2004). The therapy phase is the most suitable step where AR could contribute. The AR assembly applications combine virtual objects with the real environment to enhance the assembly design and planning process by using real time augmentation with tracking technologies and display in the worker field of view (Wiedenmaier et al. 2003). By superimposing the sequential instructions information to guide an assembly task to the workers field of view, assembly completion time and the amount of effort could be reduced (Ong and Wang 2011). When designing virtual content for sequential instructions, it is not necessary to visualize highly repetitive assembly tasks unless they are executed in different contexts or with different objects (Eversheim and Patow 2001).

Support tools, such as, paper or electronic manuals, drawings or schematics, are often detached from the equipment. Hence, the workers usually have to change their focus between the instructions and the parts being assembled, which cause loss of valuable time, increased assembly errors, repetitive motions and reduced productivity (Ong et al. 2008). AR technology allows integrating any virtual information into the exact place where the worker performs an action, which results in no change in focus of the worker. If the instructions are designed as three dimensional animations, then the worker can even observe the action from different angles to understand the involved spatial relationships (Reiners et al. 1998). AR can also promote faster learning of simple assembly tasks without referencing separate training material (Boud et al. 1999).

12.3.3 Collaborative Operations

AR technology has been also used in collaborative operations, such as, product design, collocated or remote design meetings, games, etc. In the scope of meeting scenarios, basic components necessary for a meeting are summarized by Regenbrecht et al. (2002) as; a physical place for face-to-face communication, a presentation device, necessary materials such as documents or notebooks, and intra- or inter-net access. To satisfy these real world conditions with an AR scenario, following requirements are necessary to be fulfilled; seamless access to all digital data, visualization of 3D content, interaction capability with the digital data, and remote collaboration. In terms of design, collaboration is a human centered technical activity which involves more than one designer in the process of designing a product (Lu et al. 2007). For the sake of the effectiveness of the collaborative design process, it is better that the designers have good awareness of each other together with intensive interaction among themselves (Shen et al. 2010). Experts or designers at different locations can also interact with the same product or 3D model through an information sharing environment, which can be in the form of e.g., live video (Monika et al. 2000). Since the users can see the real world while working on the virtual models, they feel more comfortable and safer (Kaufmann and Schmalstieg 2003).

Collaboration is necessary in product design and development. For example, in automotive and aerospace industry, since design and development processes consist of many iterative steps, pieces of a product have to fit with each other geometrically and functionally and be prepared for production and servicing processes. These steps involve taking into account all the requirements of these processes and finding the best solution in which a lot of decision making processes take place which cannot be done by a single person. Therefore, in most of the companies, frequent meetings happen with the specialists to make decisions. In those meetings, designers or developers use digital data (e.g., CAD models), physical mockups or prototypes, which are time consuming and expensive to construct. Moreover, users have to be desk-bound when using CAD solutions and they can hardly inspect the spatial relationships in the 3D model from a 2D monitor (Shen et al. 2010). One possible solution for this issue is presented in (Regenbrecht et al. 2002). The developed system namely MagicMeeting provides a collaborative AR system with which a group of experts discuss the design of a product. In the developed system, the digital version of the object to be designed is displayed on a plate with a marker and the experts can interact with the virtual object by using a tangible user interface. All the experts wear a HMD and see the image coming from an external camera which displays augmentations. One of the most appealing functionality of this system is that an unlimited number of virtual 2D desktops can be placed in the AR scene with which an expert can see the 2D representation of the 3D model as in the traditional 2D desktop applications and make changes on the model.

AR can offer integration of 3D model of the item into the meeting environment (Regenbrecht et al. 2005), which is easier to modify and make radical changes without constructing an actual physical model of an item. Ong and Shen (2009) presented a collaborative AR system for product design. In this system, each user wears a head-mounted device and can freely walk around the objects to observe the augmentations from different perspectives. The users can be in the same room or distributed at different locations. One advantage of this system is that using the AR technology modifications made to a part design can be displayed dynamically before the CAD model is updated. With 3D model augmentations, users can decide easily whether the updated design is what they desire. The CAD model of the object will only be updated in the server after the designer is fully satisfied with the modification. However, manipulating 3D models by using AR systems as in the traditional CAD systems is a more difficult task because of the low accuracy of tracking and registration of AR systems currently by which one may not design the model with precise parameters due to the software and hardware constraints (Shen et al. 2010).

Collaborative operations and AR technology have been also combined for educational purposes. Kaufmann and Schmalstieg (2003) proposed a 3D construction tool, namely Construct3D, for mathematics and geometry education. The aim is to enhance spatial abilities of the students and maximize transfer of learning by supporting various teacher-student interaction scenarios. The users of the system wears stereoscopic see-through HMD with camera for tracking and a back-pack computer, pinch gloves for two-handed input. The user can inspect augmented 3D objects together via a networked architecture seeing each other's actions. The system is using marker-based tracking for displaying augmentations. The results of conducted experiments shows that this kind of teaching environment has so many advantages that may change the traditional student-teacher interactions in the future.

Another important issue in these type of collocated or remote AR environments is the synchronization of the scenes on the views of the participants. The collaboration process starts with visualization, in which each user views and examines the virtual objects by using e.g., a head-mounted device. Any user is able to interact with the features of the virtual objects by using, for example, haptic devices, and other users can discuss the changes suggested. If a change is accepted in the meeting, it is applied to the object and needs to be synchronized on the view of each user (Ong and Shen 2009). A mechanism to prevent possible confusions caused by synchronization issues is proposed by Shen et al. (2010). The proposed mechanism allows a single user to modify the virtual objects only if the user has the editing right. The changes made by the user, who has the editing right, is further discussed within the group.

12.3.4 Training

In order to improve capability, productivity and performance of the users or employees, training tasks are crucial for the companies. Training operations demand high expenses for companies especially when the expensive components are to be learned or repaired by the customers or employees (Boulanger 2004). Companies invest large budgets to develop virtual models for training their employees to reduce execution time and prevent possible errors. Moreover, if the actual equipment is too complex, not all parts of the equipment will be modeled since modeling is an expensive task. This situation may increase the risk of facing surprises that may cause confusion for the employee (Raczynski and Gussmann 2004). There are several types of training depending on the context, such as, physical training, on job training, organizational training, computer-based training, etc. It is considered that the computer-based training is rather inadequate for extensive job-related tasks in aircraft systems since developing detailed instructional materials, such as, scenarios, various types of aircraft components and faults, is complex and costly (Haritos and Macchiarella 2005). Moreover, experienced technicians usually have a special way of handling problems or accomplishing tasks and subsequently have a unique way of teaching a task. These factors may not be covered in training procedures and novice technicians may be trained differently by other experts or trainers. In order to provide a way of standardizing training pro-cedures, an AR system can be developed. By using the support of AR-based solutions delivery of expert levels of information in real time to novice workers can reduce the possible errors. Moreover, instead of creating a simulation environment for training purposes, using an AR application the training can be applied in the real working environment, which saves time and budget, and also enhances the training process by enabling delivery of online feedback in real time (Young et al. 1999).

AR has been used as a support tool for training processes in various fields. Regenbrecht et al. (2005) developed an AR system for improving driver's car-handling skills, and driving experience by training the drivers for unexpected events or various environmental conditions that pose dangerous situations for the drivers. The developed system augments various scenarios as animations through head-mounted displays that both the trainer and the trainee wear within the car while driving. The virtual contents are delivered to the head-mounted displays by notebook computers placed on the back seat. The developed system was considered as a contribution to the existing applications and effective for illustrating the real scenes. In another study, a computer-based training system was developed by using benefits of AR (Pathomaree and Charoenseang 2005). In the developed system, AR is used for instructing the user with steps to assemble pieces of puzzled objects into one big square or cube. A camera is used to track positions of the objects and a desktop computer is used to calculate the correct positions for each of the objects and display instructions such as texts, objects IDs, wire frames, images to direct the user for a correct placement of the objects. The system also warns the user in case of an incorrect placement of an object. The user can see the augmentations either from the monitor or from a head-mounted display. The user tests of the system showed that incorporating AR into training saved up to 85% for the first time of assembly and 61% in the second time of assembly. It is also presented in the study that the number of assembly steps is decreased by using AR training.

Remote training operations are also an important application area of AR. Training is provided to the novice technicians on site by the remote experts. In the study of Boulanger (2004), a training practice for repairing an ATM switch was managed by a remote expert giving instructions to the trainees at different locations. The training practice involves removing a switchboard from an ATM switch and installing a chip on the switchboard. While the trainee tries to handle these oper-ations, the other trainees watch the actions of the local trainee through remote workstations to get some experience. The local trainee is equipped with a wearable computer which involves a camera and a microphone to communicate with the remote expert. In order to share the view of the local trainee with the remote expert and other trainees, a client-server architecture was established. The developed system uses vision-based tracking for overlaying augmentations by detecting markers specific for each instructions. Moreover, since the process of vision-based tracking is computationally expensive, this process was handled on the server in order to reduce the workload of the wearable computer. Once the camera on the wearable computer captures the live video and encodes it, it transfers the video to the server to analyze and calculate the positions of the augmentations related to the

Table 12.1 Comparison of references in terms of tracking and interaction technologies; Maintenance, Assembly, Collaborative operations, Training

<table>
<tr><td colspan="2" rowspan="2"></td><td colspan="4" align="center">Interaction Device</td></tr>
<tr><td align="center">Optical see-through HMD</td><td align="center">Video see-through HMD</td><td align="center">Projector</td><td align="center">Monitor Display</td></tr>
<tr><td rowspan="2">Tracking</td><td>Marker-based</td><td>(Haritos and Macchiarella, 2005), (Ong and Wang, 2011), (Reiners et. al., 1998), (Kaufmann and Schmalstieg, 2003), (Ong and Shen, 2009), (Shen et. al., 2010)</td><td>(Henderson and Feiner, 2007), (Ke et. al., 2005), (Regenbrecht et. al., 2005), (Regenbrecht et. al., 2002), (Boulanger, 2004)</td><td></td><td>(Wang et. al., 2014), (Raczynski and Gussmann, 2004)</td></tr>
<tr><td>Markerless (vision-based)</td><td>(Platonov et. al., 2006), (Wiedenmaier et. al., 2003), (Boulanger, 2004)</td><td>(Henderson and Feiner, 2009), (Boud et. al., 1999), (Regenbrecht et. al., 2005), (Regenbrecht et. al., 2005)</td><td>(Gurevich et. al., 2012)</td><td>(Rose et. al., 1995), (Pathomaree and Charoenseang, 2005), (Pathomaree and Charoenseang, 2005)</td></tr>
</table>

markers identification, and finally sends the related 3D model information to the wearable computer, which displays the model on top of the marker. The superimposed 3D model can be an instruction of pulling out a switchboard from the ATM or a pointer image that shows where the necessary action should be applied. After the 3D model is superimposed on the physical marker, the trainee performs the action and the remote expert either directs the trainee for the next step or gives further instructions in case of an incorrect action.

Overall summary of the AR researches in maintenance, assembly, collaborative operations and training is presented in Table 12.1.

12.4   Conclusion

In this chapter, we summarized current state of AR in the industrial operations. The hardware and software systems of AR have been advancing significantly fast, however, there are still some associated problems with this technology such as tracking and registration, resolution and narrow field of view among HMDs, tracking dynamic and deformed objects, high battery usage of devices etc. that are needed to be solved in order to fully adopt AR technology into the industry and also for acceptance of the people. There is also an important need for making smaller, portable, lighter and cheaper AR devices which will also enable fast growth of AR applications. However, the benefits of AR in terms of assisting people in the field with digital manuals or animations presented onto the real scene makes it one of the most crucial technologies of the next decade. AR is one of a kind technology that enables remote interactive collaboration between people, low investments on manufacturing and training, less operational errors and shorter product release time.

In order to improve current work of AR in industry, authoring tools should be developed to ease the preparation of digital content since it takes significant amount of time and investment to produce digital manuals, 3D models of the objects or other types of augmentations. Moreover, Mixed Reality (MR) should be taken into account for industrial purposes. MR adds virtual contents onto the real scene same as AR, but it anchors the virtual contents to a place in the real environment facilitating user-to-computer and user-to-user interaction and allowing them to be treated as real. For example, with Hololens device being the most advanced MR device in the current market, one can construct the virtual representation of another person remotely connected to the system and display it as an augmentation on its scene. Since the Hololens can track the position and gestures of the user, the augmentation will play the same gestures and movements of the remote person. This technology can be applied in the industry. An example use case is that while the field technician works on an automobile engine and uses a HMD to access AR content, a virtual character can be added to the scene showing the technician the demonstration of the task or giving verbal instructions etc.

References

Azuma RT (1997) A survey of augmented reality. Presence: Teleoper Virtual Environ. doi:10.1162/pres.1997.6.4.355

Azuma RT, Baillot Y, Behringer R, Feiner S, Julier S, MacIntyre B (2001) Recent advances in augmented reality. IEEE Comput Graphics Appl. doi:10.1109/38.963459

Boud AC, Haniff DJ, Baber C, Steiner SJ (1999) Virtual reality and augmented reality as a training tool for assembly tasks. In: Proceedings of IEEE international conference on information visualization, London, pp 32–36. doi:10.1109/IV.1999.781532

Boulanger P (2004) Application of augmented reality to industrial tele-training. In: Proceedings of first Canadian conference on computer and robot vision, London, ON, Canada. doi:10.1109/CCCRV.2004.1301462:320-328

Chryssolouris G (2006) Manufacturing systems theory and practice, 2nd edn. Springer, New York

Didier JY, Roussel D, Mallem M, Otmane S, Naudet S, Pham Q, Bourgeois S, M'egard C, Leroux C, Hocquard A (2005) AMRA: Augmented reality assistance for train maintenance tasks. Workshop industrial augmented reality, 4th ACM/IEEE ISMAR, Vienna, Austria Oct 2005

Dini G, Mura MD (2015) Application of augmented reality techniques in through-life engineering services. In: Procedia CIRP, vol 38, pp 14–23, ISSN 2212-8271

Eversheim W, Patow C (2001) Assembly structure and work place design in single piece and small series production. In: Landau K, Luczak H (eds) Ergonomie und Organisation in der Montage, Mnchen, Germany, pp 581–612

Gurevich P, Lanir J, Cohen B, Stone R (2012) TeleAdvisor: a versatile augmented reality tool for remote assistance. In: Proceedings of the SIGCHI conference on human factors in computing systems (CHI '12). ACM, New York, NY, USA, pp 619–622. doi:10.1145/2207676.2207763

Haritos T, Macchiarella ND (2005) A mobile application of augmented reality for aerospace maintenance training. In: 24th digital avionics systems conference, vol 1. doi:10.1109/DASC.2005.1563376

Henderson SJ, Feiner S (2007) Augmented reality for maintenance and repair (ARMAR). Technical report AFRL-RH-WP-TR-2007-0112, United States Air Force Research Lab, July 2007

Henderson SJ, Feiner S (2009) Evaluating the benefits of augmented reality for task localization in maintenance of an armored personnel carrier turret. Paper presented at the 8th IEEE ISMAR '09, IEEE computer society, Washington, DC, USA, pp 135–144

Kaufmann H, Schmalstieg D (2003) Mathematics and geometry education with collaborative augmented reality. Comput Graph 27(3): 339–345. doi:10.1016/S0097-8493(03)00028-1

Ke C, Bo K, Chen D, Li X (2005) An augmented reality-based application for equipment maintenance. In: Proceedings of affective computing and intelligent interaction: first international conference (ACII), Beijing, China, 22–24 October, pp 836–841. doi:10.100711573548 107

Li GY, Xi N, Yu M, Fung WK (2004) Development of augmented reality system for AFM based nanomanipulation. IEEE-ASME Trans Mech 9(2):358365

Lu SCY, Elmaraghy W, Schuh G, Wilhelm R (2007) A scientific foundation of collaborative engineering. Ann CIRP 56(2):605–634

Lukosch S, Billinghurst M, Alem L, Kiyokawa K (2015) Collaboration in augmented reality. J Comput Support Coop Work (CSCW) 24(6):515–525. ISSN="1573-7551"

Monika B, Michael C, Kaj G, Peter K, Preben M, Dan S, Peter O (2000) Collaborative augmented reality environments: integrating VR, working materials, and distributed work spaces. In: Proceedings of the 3rd international conference on collaborative virtual environments, CA, USA, pp 47–56

Nee AYC, Ong SK (2013) Virtual and augmented reality applications in manufacturing. In: IFAC proceedings volumes, vol 46, pp 15–26. ISSN 1474-6670

Nee AYC, Ong SK, Chryssolouris G, Mourtzis D (2012) Augmented reality applications in design and manufacturing, CIRP Ann—Manuf Technol 61(2): 657–679. ISSN 0007-8506

Nevins JL, Whitney DE (1980) Assembly research. Automatica 16(6):595–613. doi:10.1016/0005-1098(80)90003-5

Oliveira R, Farinha T, Raposo H, Pires N (2014) Augmented reality and the future of maintenance. In: Proceedings of maintenance performance measurement and management (MPMM) conference. doi: 10.14195/978-972-8954-42-0 12:81-88

Oliveira R, Farinha T, Singh S, Diego G (2013) An augmented reality application to support maintenance is it possible? In: (MPMM) Maintenance performance measurement and management, Lappeenranta, Finland, pp 260– 271

Ong SK, Nee AYC (2004) Virtual and augmented reality applications in manufacturing. Springer, London. ISBN 1-85233-796-6

Ong SK, Shen Y (2009) A mixed reality environment for collaborative product design and development. CIRP Ann—Manuf Technol 58(1):139–142. ISSN 0007-8506, doi:10.1016/j.cirp.2009.03.020

Ong SK, Yuan ML, Nee AYC (2008) Augmented reality applications in manufacturing: a survey. Int J Prod Res 46(10):2707–2742. doi:10.1080/00207540601064773

Ong SK, Wang ZB (2011) Augmented assembly technologies based on 3D barehand interaction. CIRP Ann—Manuf Technol 60(1):1–4. ISSN 0007-8506, doi:10.1016/j.cirp.2011.03.001

Pathomaree N, Charoenseang S (2005) Augmented reality for skill transfer in assembly task. IEEE international workshop on robot and human interactive communication, pp 500–504. doi: 10. 1109/ROMAN.2005.1513829

Platonov J, Heibel H, Meier P, Grollmann B (2006) A mobile markerless AR system for maintenance and repair. In: IEEE/ACM international symposium on mixed and augmented reality, Santa Barbard, CA, pp 105–108. doi: 10.1109/ISMAR.2006.297800

Raczynski A, Gussmann P (2004) Services and training through augmented reality. In: 1st European conference on visual media production (CVMP), pp 263–271

Regenbrecht H, Baratoff G, Wilke W (2005) Augmented reality projects in the automotive and aerospace industries. IEEE Comput Graphics Appl. doi:10.1109/MCG.2005.124

Regenbrecht H, Wagner M, Baratoff G (2002) Virtual reality 6: 151. doi:10.1007/s100550200016

Reiners D, Stricker D, Klinker G, Müller S (1998) Augmented reality for construction tasks: doorlock assembly. In: Proceedings of the international workshop on augmented reality: placing artificial objects in real scenes: placing artificial objects in real scenes (IWAR '98), USA, pp 31–46

Rose E, Breen D, Ahlers KH, Crampton C, Tuceryan M, Whitaker R, Greer D (1995) Annotating real-world objects using augmented reality. In: Rae E, John V (eds) Computer graphics. Academic Press Ltd., London, UK, pp 357–370

Shen Y, Ong SK, Nee AYC, (2010) Augmented reality for collaborative product design and development, Design Studies, doi:10.1016/j.destud.2009.11.001, 31(2):118–145

Wang J, Feng Y, Zeng C, Li S (2014) An augmented reality based system for remote collaborative maintenance instruction of complex products. In: IEEE international conference on automation science and engineering (CASE), Taipei, pp 309–314. doi: 10.1109/CoASE.2014.6899343

Wiedenmaier S, Oehme O, Schmidt L, Luczak H (2003) Augmented reality (AR) for assembly processes design and experimental evaluation. Int J Hum Comput Interact. doi:10.1207/S15327590IJHC1603

Young AL, Stedman AW, Cook CA (1999) The potential of augmented reality technology for training support systems. In: International conference on human interfaces in control rooms, cockpits and command centres, pp 242–246. doi: 10.1049/cp:19990194

Chapter 13
Additive Manufacturing Technologies and Applications

Omer Faruk Beyca, Gulsah Hancerliogullari and Ibrahim Yazici

Abstract Additive Manufacturing (AM) Technologies is has become widely popular manufacturing technique in the last 30 years. Additive manufacturing uses 3D Computer Aided Design (CAD) model to manufacture parts layer by layer through adding materials. Additive Manufacturing allows to manufacture complex geometries that cannot be manufactured using conventional manufacturing techniques. Additive manufacturing permits customized designs and prototypes to be produced easily. Over the past three decades thorough research has been done to commercialize AM techniques in varied areas. As a result of these research, AM techniques are being used in automotive, aerospace, biomedical, medicine, energy and in many other areas. Due to its advantage over the traditional manufacturing techniques, additive manufacturing is seen as one of the enablers that started fourth industrial revolution. In this chapter, we will review additive manufacturing processes by introducing its history, presenting different technologies of AM, giving examples of AM usage in different application areas and its impact on our society.

Abbreviations

AM	Additive Manufacturing
CAD	Computer Aided Design
STL	Standard Tessellation Language
SL	Stereolithography
3DP	Three Dimensional Printing
SLS	Selective Laser Sintering
LOM	Laminated Object Manufacturing
LENS	Laser Engineered Net Shaping
LCA	Life-Cycle Analysis
EIA	Environmental Impact Assessment
JIT	Just in Time

O.F. Beyca (✉) · G. Hancerliogullari · I. Yazici
Istanbul Technical University, 34367 Macka, Sisli, Istanbul, Turkey
e-mail: beyca@itu.edu.tr

© Springer International Publishing Switzerland 2018   217
A. Ustundag and E. Cevikcan, *Industry 4.0: Managing The Digital Transformation*,
Springer Series in Advanced Manufacturing, https://doi.org/10.1007/978-3-319-57870-5_13

13.1 Introduction

Up to now, prevalent method in manufacturing industry has been subtractive methods, which they generate a product with high precision by grinding a material or cutting away it from the block of solid material. It can be done by either manually or automatically. Most important aspect of subtractive methods, especially automatically one, is to generate exact shapes with high precision. However, there is another manufacturing method that may challenge traditional subtractive methods; additive manufacturing or it may also be called 3D printing. Additive manufacturing technique builds a product by adding layers onto other layers using raw material.

Additive manufacturing's roots may be traced back to nineteenth century, and several researchers have made different definitions of additive manufacturing in time. The roots of AM date back to nineteenth century in the area of topography and photosculpture. In 1972, Ciraud's defined additive manufacturing as to building a product by melting materials using a beam of energy and laying melted materials on top of layer. In the meantime, scientist Hideo Kodama and Alan Herbert developed relevant technologies with additive manufacturing and tested them at the beginning of 1980s. This technology is described as creating prototypes by adding materials layer by layer. In 1986, Chuck Hull invented stereolitography (SL) technique that can be recognized as milestone in the history of additive manufacturing. Charles Hull first published his research on stereolthography machine in 1984 then patented it in 1986. Other technologies in additive manufacturing such as Selective Laser Sintering (SLS) and Fused Deposition Modeling (FDM) were introduced in the late 1980s. With the advancing technology, several AM companies are established in mid 1990s.

For more than two decades AM technology and techniques are developed to answer industrial needs. In 2005, RepRap (Replication Rapid-Prototyper) project was introduced. RepRap project enabled individuals to utilize 3D printing by providing open source codes. It was first introduced by Dr. Adrian Bowyer from University of Bath, England (Matias and Rao 2015). RepRaps have the ability to produce almost all of its parts, in other words it can produce itself. The project assisted the technology to gain importance and revolutionize rapid prototyping, tinkering and personal manufacturing. Since the invention of additive manufacturing, the AM technology has been used in many industrial areas such as automotive, manufacturing, aviation, medical, jewelry.

13.2 Additive Manufacturing (AM) Technologies

Over the past 30+ years researchers and industrial companies has developed several techniques for additive manufacturing and rapid prototyping. We will review some of these techniques in the following sub-sections. A typical AM technology involves three phases:

1. A 3D solid model is designed in a CAD software and converted into a Standard Tessellation Language (STL) format (Kumar and Dutta 1997) or other newer AM file format (Lipson 2013).
2. AM machine manipulates the file to adjust the position and orientation of the part.
3. AM machine manufactures the part by sequential layers (Huang et al. 2013).

13.2.1 Stereolithography

Stereolithography (SL) is the first developed rapid prototyping technique patented by Charles (Chuck) W. Hull in 1986. Then 3D Sytems Inc is founded to commercialize his patent. Working principle of SL is to solidify the photosensitive liquid polymer by using an ultraviolet laser.

3D modeled part in CAD software converted into STL file. In STL file, modelled part is divided into cross sections, which contains information about each layer. The thickness of the layer defines the resolution, which is dependent on the equipment used. A support structure is needed to build to part on. A laser beam is traces the cross- section information provided by STL file to build the part layer by layer. When process is completed excess liquid polymer is drained saved for reuse (Chua et al. 2010; Cooper 2001; Kruth 1991; Wong and Hernandez 2012).

13.2.2 3DP

Three dimensional printing (3DP) is patented in 1993. In 3DP technique, polymer powder is fused using water based liquid binder supplied from jet nozzle. Powder particles are glued together layer by layer. Wide range of polymers can be used for this process. Advantages of this process are; it is fast and material cost is low. On the other hand, surface finish is rough and part size is limited.

13.2.3 Fused Deposition Modeling

In Fused Deposition Modelling (FDM) technique, print head liquefies the thermoplastic material and deposits in layers on substrate. The thickness of the layers are usually 0.25 mm. The FDM technique is patented in 1992. The description of the technique was first introduced by Crump (1991). The materials is heated just enough to make it viscous so that it can solidify immediately. Different types of materials can be used such as, wax, metals and ceramics. Low maintenance cost and its compact size are the positive attributes of FDM technique. On the other hand temperature fluctuations results in delamination and printing time is long (Skelton 2008).

13.2.4 Selective Laser Sintering

Selective Laser Sintering (SLS) technique is a process, where printing material (polymers, metals, ceramics and glass) are fused together by using carbon dioxide laser beam (Kruth et al. 2005). The technology is patented in 1989 (Deckard and Beaman 1988). The powder material is fused layer by layer according to specified design. The sintered powdered material will construct the 3D design while the un-sintered will be cleaned for reuse after the part is built. SLS is used to 3D print complex parts quickly and provide better durability compare to other additive manufacturing technologies. Disadvantage of SLS is that many parameters are needed to be tuned in order to get better surface finish (Kamrani and Nasr 2010).

13.2.5 Laminated Object Manufacturing

Laminated Object Manufacturing (LOM) technique is developed in 1988 (Feygin and Hsieh 1991). LOM is the combination of subtractive and additive manufacturing techniques. In LOM process, a sheet of material represents each layer. Laser cutting is used to manufacture each sheet of material according to STL file information. Later sheets of materials are attached together by using heat and pressure (Noorani 2006). Materials such as paper, plastics, metals and composite materials can be used as building material. LOM is an inexpensive additive manufacturing technique but it can create internal cavities, which can decrease the quality of the products. LOM also causes waste by subtracting material (Kamrani and Nasr 2010).

13.2.6 Laser Engineered Net Shaping

Laser Engineered Net Shaping (LENS) is an additive manufacturing process where melted metal powder is injected to a specific location to build the 3D part. The technology is developed at Sandia National Labs and patented in 2000 (Mudge and Wald 2007). In addition to building 3D parts LENS can also be utilized to repair damaged parts (Griffith et al. 1999).

13.2.7 Advantages of Additive Manufacturing

Some of the advantages that provides superiority to additive manufacturing over the conventional manufacturing techniques are listed below:

- *Decrease in waste*: In traditional subtractive manufacturing techniques, large amount of raw material is cut and wasted in order to manufacture a part. On the other hand, AM only uses the raw material needed to manufacture the part and also leftover raw material can be reused.

- *Increase in time*: With conventional techniques, producing a prototype may take several days or even weeks while AM technologies can provide functional working prototype mostly within hours.
- *Production flexibility*: Most of the parts need several manufacturing steps and assembly workshops which affects the overall quality of the product. AM technique manufacture the part in one process and eliminates the effect of operator on product quality.
- *Increase in variety*: Since complex parts can be produced with little setup cost (changing design in CAD software) custom designed products can easily be produced with little cost.
- *Few constraints*: Anything that can be designed in a CAD software can be produced.
- *Decrease in labor skill*: Automated production requires little or no operator skill.

13.2.8   Disadvantages of Additive Manufacturing

Although AM technology is rapidly growing industry, some of its drawbacks prevent AM to compete with traditional manufacturing techniques. We can list some of these challenges in the following:

- *Dimensions limitation*: In most of the AM processes, liquid polymers or powder materials are used to manufacture layers. Due to these material's low strength large size parts cannot be produced using AM processes.
- *Rough Surface Finish*: Producing the part layer by layer often results into imperfect surface finish and frequently finished parts need postproduction.
- *Mass Production*: AM processes is not suitable for mass production considering its low speed for producing one part.

13.3   Application Areas of Additive Manufacturing

Advantages mentioned in the previous sections makes AM crucial for industries in competitive environment. In addition, researches constantly working on refining the AM process to eliminate the drawbacks. AM technologies are being used in wide variety of application areas, such as aerospace, automobile, medicine, energy, etc.

Wohlers report (Wohlers and Caffrey 2013) in 2013 reveals that organizations mainly uses AM for functional parts and prototypes for fit and assembly. The survey asks the question "How do your customers use the parts built on your AM systems?" with the following answering options:

- Visual aids (for engineers, designers, medical professionals, etc.)
- Presentation models (including architectural)
- Prototypes for fit and assembly
- Patterns for prototype tooling (including silicone rubber molds)
- Patterns for metal castings
- Tooling components (created directly on an AM system)
- Functional parts (for short run, series production, prototyping, etc.)
- Education/research
- Other.

The following figure shows the other usage purposes' percentages (Fig. 13.1).

Wohlers report indicate that AM technologies is especially fit for low volume of complex products such as functional prototypes. We are going to review some of these applications in different areas in the following sections.

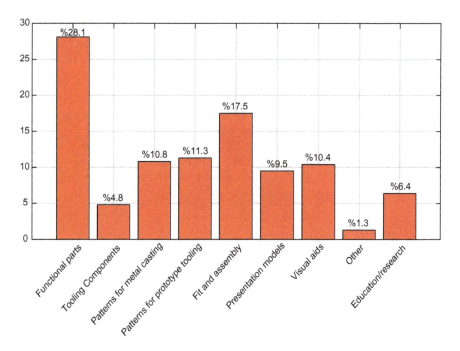

Fig. 13.1 Uses of additive manufacturing in industry [Adapted from (Wohlers and Caffrey 2013)]

13.3.1 Medical

Although rapid prototyping has been used for a long time in the manufacturing industries, it has just started to take place in health care applications. In this sense, in various medical applications, such as individual patient care, research and as an educational and training tool rapid prototyping is brought in for the last years.

13.3.2 Surgical Planning

In the surgical area, for improving the understanding of complexities, rapid prototyping has been involved, making the diagnostic quality and pre-surgical planning more developed. Moreover, its utility in craniofacial surgery, maxillofacial surgery (Wagner et al. 2004), pelvic surgery (Guarino et al. 2007; Hurson et al. 2007) neurosurgery, spine surgery (Paiva et al. 2007), cardiovascular surgery (Armillotta et al. 2007; Kim et al. 2008) and visceral surgery (D'Urso et al. 1999) is proven by the studies in the listed areas, since it improved not only the diagnosis stage but also the treatment as a consequence of the advanced and accurate pre-planning (Mavili et al. 2007) which portrays the surgical steps beforehand and foreseeing the possible complications by the 3d assessment. Consequently, in addition to the reduction in the operating time and the use of operating rooms in a cost effective way, it also improves the process of radiotherapy planning (Kalet et al. 1999; Sun and Wu 2004) and producing radiation shields for individuals (Zemnick et al. 2007).

13.3.3 Implant and Tissue Designing

The rapid prototyping technique is additionally used in medical prosthesis and designing implants allowing the prostheses to be customized, unlike the standard sized ones which are available for complying with the requirements of many surgical procedures, but not all of them. In this sense, the need of implants that are customized lies under the presence of patients who are not in the standard range in terms of the diseases they have or the size of the implants they need, and more successful surgical outcomes consequently with the fitting and meeting the needs of the body.

The technique of rapid prototyping has been involved in reconstructing many anatomical structures including mandible (D'Urso et al. 2000; Singare et al. 2007) and dental restoration (Lee et al. 2008) in facial surgery and others such as hip (Dai et al. 2007), femoral (Harrysson et al. 2007), hemi-knee joint reconstruction (He et al. 2006; Wang et al. 2004) in which bioceramics, biocompatible metals and ceramics are preferred in place of bones.

In addition to this technique's benefits in reconstructing bones, the rapid prototyping can be used for soft tissue replacement and creating tissue scaffolds for the growth of cells as well (Boland et al. 2006; Campbell and Weis 2007), making the future application of producing artificial organs for the patients possible with further research.

13.3.4 Medical Research

Rapid prototyping also creates new areas for scientific researches. As it is clear, research which is based on phantoms rather than objects produced by rapid prototyping will only make not a satisfactory level of understanding available for both physiological processes (Canstein et al. 2008; Chung et al. 2006) and complex pathologies (Kim et al. 2008; Tek et al. 2008). Because the 3d solid models will portray the complexities better than visuals which are 2d or 3d. In this regard, various types of materials such as silicon or metals can be chosen to be used for producing different models according to the needs of the patients in the areas of hemodynamic or aerodynamic (Giesel et al. 2009).

13.3.4.1 Medical Education and Training

Procedures in surgery require human anatomy and relations of these different structures to be known completely. For this reason, the medical schools try to give the detailed knowledge to their students by the human cadavers in the first stage, and then intensify it with real surgery for the students to acquire wisdom, as gaining experience in the area that will be chosen later, is desired before dealing with real patients, since training of these procedures beforehand will improve not only the abilities of the surgeon (Knox et al. 2005), but also the results without any risk of complications. In order to gain this knowledge, although 2d or 3d visualizations are widely used, they are generally not enough for creating a sufficient understanding of the humans' anatomical structures (Suzuki et al. 2004). Because of that, rapid prototyping is introduced for enhancing their learning. Both the young or experienced surgeons, after efficiently training on models rather than operating the patients first, feel more confidence for the real surgery as they had the chance of practicing various surgical steps to decide whether one of them is better than other treating methods or not, in order to determine their strategy for the actual surgery (Mavili et al. 2007).

13.3.5 Automotive

CRP Technology (Italy) to produce motorsports components has successfully applied various AM techniques. Some of the produced parts using AM techniques by CRP Technology are; MotoGP 250R air boxes, camshaft covers for MotoGP engines, F1 gearboxes, motorbike supports and dashboards (CRPTechnology). AM techniques provides severela advantages in manufacturing MotoGP parts. For instance, AM techniques decreases the weight of F1 gearbox by 20%–25% and volume by 20%. In addition, ability to manufacture complex design provides twice in torsion stiffness, less power absorption, and less gear wear (Guo and Leu 2013).

Optomec utilizes LENS technique to produce components of Red Bull Racing car, such as drive shaft spiders and suspension mountain brackets made up of Ti6A14 V (titanium alloy) (Optomec). Using AM technique, LENS, provides more than 90% raw material saving and decreases the production time and cost considerably(Guo and Leu 2013).

Prometal used a modified version of 3DP technique, Prometal process, to build components of passenger and race car engines, such as cylinder heads, engine blocks, and intake manifolds. Using AM technique decreases the production time of car engines significantly. For instance, an engine block with cooling passages and oil recirculation lines can be manufactured fully in just one week (Guo and Leu 2013).

13.3.6 Aerospace

One of the assembling divisions with high prospects for 3D printing is aerospace production industry. The potential for fuel investment funds due to much more lighter parts fabricated through 3D printing is the most appealing benefit for the avionic business (Joshi and Sheikh 2015). Moreover, generation in aviation can possibly diminish decommissioning-related CO_2 discharges and total primary energy supply (TPES) requests (Gebler et al. 2014).

This industry works around 2 essential guideline necessities—low weight and high wellbeing. 3D printing has possessed the capacity to help lessening in weight through unpredictable and net shape producing with less number of joints and many-sided geometry. However, from the safety viewpoint, it is as yet far before being the solid standard. Many difficulties, for example, printing designs, porosity developed, and uneven print flow, should be unraveled and dispensed with totally. It is simply an issue of time. Once that happens, 3D printing would supplant increasingly conventional assembling methods right now utilized as a part of the avionic business and will definitely have a managed adjustment and development (Joshi and Sheikh 2015).

Among the various additive manufacturing procedures, the ones that meet the airplane business necessities are Selective Laser Sintering (SLS), Selective Laser Melting (SLM), Electron Beam Melting (EBM), and Wire and Arc Additive Manufacturing (WAAM) (Kobryn et al. 2006; Lyons 2014; Wood 2009).

A considerable measure of research is going ahead into utilizing distinctive sorts of metals and metal amalgams for additive manufacturing. Polymers, fired composites, compounds of aluminum, steel, and titanium articles can be printed with a base layer thickness of 20–100 μm, contingent upon the additive manufacturing strategy utilized and the physical condition of the material (Hopkinson et al. 2006). However, from the aeronautic trade perspective, more significance is given towards Ti-and Ni-based amalgams (Campbell and Weiss 2007; Uriondo et al. 2015). Nickel-based composites favored in aviation because of their malleable properties, harm resilience, and consumption/oxidation resistance (Nijdam 2010). On account of Ti-Combination, Ti-6Al-4 V, as the cooling rate is diverse for SLM and EBM forms, the microstructure of the resultant Ti6Al4 V segments is distinctive bringing about various hardness and malleability for these two procedures (Ding et al. 2002).

Though progressing serious research works, the reception rate of 3D printing in the aviation producing industry is still moderate. This can be ascribed to a few reasons. The essential concern is the strict certification necessities that are inalienable to the security of air and space creates. Testing and security measures for additive manufacturing in aviation are still a work in progress. Additionally, it is difficult to distinguish an arrangement of certification principles given that distinctive additive manufacturing innovations are still on their approach to being completely developed (Joshi and Sheikh 2015).

13.3.7 Education

Educational foundations at all levels are starting to perceive the estimation of 3D printing innovation and have started to fuse these machines into their research facilities. A 3D printer encourages the intelligent guideline in specialized ideas and frameworks reliable with the country's attention on science, technology, engineering, and mathematics (STEM) learning activities (Foroohar and Saporito 2013). The 3D printer enables a teacher to deliver physical models that learners can touch, feel, and eventually test under various physical limitations. For instance, a class could print extensions and test the differential load-bearing characteristics of various basic outlines (Martin et al. 2014).

A 3D printer can make an extensive variety of straightforward machines like apparatuses or pulleys and even screws. Particulars of apparatuses, pulleys, and coordinated frameworks of different riggings or numerous pulleys can be talked about in class before they are delivered. This procedure gives learners a more reasonable perspective of the outline and generation periods of assembling in the event that done accurately, the procedure can start learners' scholarly interest by making reckoning between the time the parts are examined and the time they really show up. At that point the arrangement of apparatuses or pulleys can be collected and put to use in classroom lab exercises (Martin et al. 2014).

The 3D printer can likewise improve classes that are based around 3D displaying programming. Most 3D demonstrating classes are instructed in PC labs and seldom result in the production of the physical models. This can prompt a detachment between manifestations in an advanced domain with imperatives experienced in the physical world. On the off chance that segments are just outlined in 3D displaying programming, learners don't perceive the entanglements that emerge when transforming those models into a physical part. A 3D printer can create exceptionally outlined parts that an understudy can physically investigate. This review will immediately affect the general plan handle. Learners will have the capacity to see their slip-ups in the part, make modifications in their advanced models, and print out another part to check that the adjustments they made to the model are adequate (Martin et al. 2014).

Innovation and designing instruction offices draw in students in a wide assortment of learning exercises. Regardless of whether students are finding out about mechanical frameworks, item outline and manufacture, or electrical frameworks, the adaptability of a 3D printer can upgrade these exercises. Electrical frameworks typically require some kind of work board or special lodging. Pre-assembled lodgings for tasks may be elusive, yet the 3D printer can be utilized to create them. The 3D printer can deliver modified nonconductive sheets or lodging for electrical circuits. Basic venture sheets for arrangement and parallel circuit undertakings can without much of a stretch be made with a 3D printer (Martin et al. 2014).

The printer is particularly valuable for more unpredictable activities. At Illinois State College, a group of students are using a 3D printer to deliver the greater part of the custom lodging segments for sunlight based controlled stereos. 3D printing can reform innovation and designing instruction. The idea of "think globally, produce locally" has never been more obvious. 3D printing innovation has made huge headways in the course of recent years and now is more reasonable than any other time in recent memory. The flexibility of this machine gives innovation and designing instructors with the capacity to draw in their students with various STEM-based exercises that assistance to meet instructive guidelines (Martin et al. 2014).

13.3.8   Biotechnology

3D printing has meaningfully affected restorative imaging in the field of legal science, taking into account anatomically revise entertainment of substantial wounds from CT and X-ray examines. For instance, models of both inside and outer injuries have been reproduced that consider better clarification of legal findings, while keeping away from the need to introduce exasperating proof within the sight of casualties' relatives (Ebert et al. 2011). 3D printing systems were utilized to reproduce skull sections from a limit compel head damage and help in weapon identification and assurance of the component of harm prompting death (Woźniak et al. 2012). A comparable utilization of 3D printing saw the amusement of a skull after a traumatic harm to derive the reason for damage, with results tantamount to those accomplished utilizing customary techniques to detach bone from the casualty (Kettner et al. 2014). Measurable appraisal of a distorted skull from the eighteenth century yielded a facial reproduction in light of a 3D printed rendition of the skull, from which creators surmised the reason for twisting (Klepáček and Malá 2012).

13.3.9   Electronics

There are tremendous applications concerning the reconciliation of 3D printing and hardware, and keeping in mind that this association is still in its early stages, the establishment that has been laid so far is promising for future attempts. A lithium particle battery has been 3D printed with suggestions in vitality storage (Sun et al. 2013). 3D printing innovation has been utilized to create a usable electrochemical cell (Symes et al. 2012). A model of an electrically conductive model was made by 3D printing a mortar based structure that contained carbon nanofibers (Czyżewski et al. 2009). Exact inkjet based 3D printing of conductive copper has been accomplished, with low expenses and low material waste, and has applications in specifically applying conductive material for circuit board generation (Gross et al. 2014).

13.3.10   Design

3D Printing models are utilized for representation apparatuses in design (Maslowski and Heise 2002). Nonetheless, they report troubles in replicating luxurious subtle elements and unsupported structures, for example, stacks and railings. Their answer was to fabricate little tops around this detail to guarantee that it survives the de-powdering stage. Building a ground plane additionally fortified the structure and ensure detail at ground level.

Another issue with design models is that 3D models of structures cannot be consistently scaled so they are sufficiently little for the manufacture volume of standard 3D Printers. The divider thickness will be too little. This implies computer aided design models must be extraordinarily intended for the prototyping stage. This is ineffective and there is the additional hazard in the human interpretation of the first computer aided design information into the model computer aided design information that data would be lost or exchanged erroneously (De Beer 2006).

13.3.11   Oceanography

Due to lower costs and expanded abilities of 3D printing advances, remarkable open doors in the realm of oceanography research are being made as well. A few illustrations incorporate 3D printed parts being utilized in self-sufficient submerged (or surface) vehicles; 3D printed copies of marine life forms being utilized to consider biomechanics, hydrodynamics, and headway; and 3D printed coral reef imitations being utilized to reestablish harmed coral reefs (Mohammed 2016).

13.4 Impact of Additive Manufacturing Techniques on Society

Despite its drawbacks, AM technology advances rapidly and plays increasingly important role in manufacturing and in our daily life.

13.4.1 Impact on Healthcare

AM technologies are being used to provide high quality and cheap healthcare that is custom-made to target specific needs of patients. This tailored approach spans from long-term personalized care of elderly (Beattie 1998) to optimize the course of theory using patients historical biological data (Ely 2009).

AM technique can be utilized to produce personalized surgical implants. Surgical implants manufactured by AM processes are reported as accurate, functions well, and aesthetically appealing (Singare et al. 2004). Additionally, AM can shorten the design and manufacturing cycle as well as decreases the delivery lead time (He et al. 2006b). Applications of implants can be listed as skull (Singare et al. 2004, 2006), elbow (Truscott et al. 2007), joint (He et al. 2006a), hip joint (Popov and Onuh 2009), and dentistry (Liu et al. 2006).

Another application area of AM process in healthcare is to manufacture tailored safety equipment and protective garments. This custom-made equipment provide excellent protection and increase performance of the user.

Although not commercialized yet, researches on tissue engineering and drug delivery devices shows great potential. One of these promising projects is CUSTOM-FIT, which aims to construct a system that will integrate design, production, and supply of personalized medical products (Gerrits et al. 2006).

13.4.2 Impact on Environment

In order to analyze the impact of any manufacturing process on environment we need to look into process time, energy utilization, primary flow of work-piece materials, and secondary flows of catalysts (Munoz and Sheng 1995; Sheng and Munoz 1993; Sheng et al. 1995). In literature, there are two main valuation methods to measure the impact of an process on environment, namely, life-cycle analysis (LCA) and environmental impact assessment (EIA) (Huang et al. 2013).

Since AM processes uses just enough raw material for building up parts, life-cycle material mass and energy consumed is reduced comparing to traditional subtractive manufacturing processes. Cutting fluids used in subtractive manufacturing processes are the main source of hazard. AM processes does not use cutting fluids reducing the environmental hazard waste (Huang et al. 2013).

13.4.3 Impact on Manufacturing and Supply Chain

In a traditional supply chain, materials move from suppliers to customers, while funds and information moves backward from customers to suppliers. Reeves (2008) reported that AM processes can decrease the number of chains in a traditional supply chain. This can be achieved by:

1. to reduce the components of a product by redesigning.
2. to place manufacturing areas near the customer.

The aforementioned innovation brought by AM processes will reduce the need of warehousing, packaging, and transportation (Huang et al. 2013).

AM technology is ideal for lean supply chain and Just in Time (JIT) manufacturing. With the help of AM, setup and changeover time will be reduced also assembly process will be shortened. This will eliminate most of the non-value added processes such as material handling and inventory management (Tuck et al. 2006) which will minimize the supply chain cost. In addition, a responsive supply chain can be constructed using AM process. A made-to-order strategy can be applied to eliminate the stock out and inventory. Customization of the products will ensure the made-to-order strategy will be implemented successfully and increase the responsiveness.

Supply chain of the spare parts faces significant challenges while giving fast maintenance and repair services. Most of the spare parts are infrequently needed but in order to provide reliable maintenance services spare parts must be kept in stock which will increase inventory and delivery cost. Another challenge is the batch size. In order the benefit from economy of scale parts are needed to be produced in large quantities, which will increase inventory cost. AM technology can be a feasible solution to this problem by producing spare parts when and in where they are needed (Walter et al. 2004).

13.5 Conclusion

In 30 years AM has become a viable manufacturing process from just a rapid prototyping technique. Several companies are using AM processes for commercial purposes. Several researches are still resuming in perfecting the AM process, which will have great impact on society in near future.

Even today AM has great impact on society in many aspects. It revolutionaries the rapid prototyping making the design of new products easier. Customized products can easily be manufactured using AM, which will improve health and quality of life. Brand new technologies such as tissue engineering will help prevent deaths from organ failing. Decrease cost in producing healthcare products will make quality life to be affordable. AM processes reduce waste and does not use hazardous materials such as cooling liquids, and thus reduce pollution. AM will simply supply chain and enable JIT manufacturing, which will greatly reduce the material handling and inventory costs.

References

Armillotta A, Bonhoeffer P, Dubini G, Ferragina S, Migliavacca F, Sala G, Schievano S (2007) Use of rapid prototyping models in the planning of percutaneous pulmonary valved stent implantation. Proc Inst Mech Eng [H] 221(4):407–416

Beattie W (1998) Current challenges to providing personalized care in a long term care facility. Leadersh Health Serv 11(2):1–5

Boland T, Xu T, Damon B, Cui X (2006) Application of inkjet printing to tissue engineering. Biotechnol J 1:910–917

Campbell PG, Weiss LE (2007) Tissue engineering with the aid of inkjet printers. Expert Opin Biol Ther 7(8):1123–1127

Canstein C, Cachot P, Faust A, Stalder A, Bock J, Frydrychowicz A, Kuffer J, Hennig J, Markl M (2008) 3D MR flow analysis in realistic rapid-prototyping model systems of the thoracic aorta: comparison with in vivo data and computational fluid dynamics in identical vessel geometries. Magn Reson Med 59:535–546

Chua CK, Leong KF, Lim CS (2010) Rapid prototyping: principles and applications. World Scientific, Singapore

Chung S, Son Y, Shin S, Kim S (2006) Nasal airflow during respiratory cycle. Am J Rhinol 20:379–384

Cooper K (2001) Rapid prototyping technology: selection and application. CRC Press, Boca Raton CRP Technology

Crump SS (1991) Fast, precise, safe prototypes with FDM. ASME, PED 50:53–60

Czyżewski J, Burzyński P, Gaweł K, Meisner J (2009) Rapid prototyping of electrically conductive components using 3D printing technology. J Mater Process Technol 209(12): 5281–5285

Dai K, Yan M, Zhu Z, Sun Y (2007) Computer-aided custom-made hemipelvic prosthesis used in extensive pelvic lesions. J Arthroplasty 22:981–986

D'Urso PS, Barker TM, Earwaker WJ, Bruce LJ, Atkinson RL, Lanigan MW, Effeney DJ (1999) Stereolithographic biomodelling in cranio-maxillofacial surgery: a prospective trial. J Cranio-maxillofac Surg 27(1):30–37

D'Urso PS, Effeney DJ, Earwaker WJ, Barker TM, Redmond MJ, Thompson RG, Tomlinson FH (2000) Custom cranioplasty using stereolithography and acrylic. Br J Plast Surg 53(3):200–204

De Beer N (2006) Advances in three dimensional printing-state of the art and future perspectives. J New Gener Sci 4(1):21–49

Deckard C, Beaman J (1988) Process and control issues in selective laser sintering. ASME Prod Eng Div (Publication) PED 33(33):191–197

Ding R, Guo ZX, Wilson A (2002) Microstructural evolution of a Ti–6Al–4V alloy during thermomechanical processing. Mater Sci Eng A, 327:233–245.

Ebert LC, Thali MJ, Ross S (2011) Getting in touch—3D printing in forensic imaging. Forensic Sci Int 211(1):e1–e6

Ely S (2009) Personalized medicine: individualized care of cancer patients. Transl Res 154(6): 303–308

Feygin M, Hsieh B (1991) Laminated object manufacturing: a simpler process. In: Paper presented at the solid freeform fabrication symposium, Austin, TX, August

Foroohar R, Saporito B (2013) Made in the USA. Time, Inc 181(15). Retrieved from http://ehis. ebscohost.com

Gebler M, Uiterkamp AJS, Visser C (2014) A global sustainability perspective on 3D printing technologies. Energy Policy 74:158–167

Gerrits A, Jones CL, Valero R, Dolinsek S (2006) A concept of manufacturing system enabling the creation of custom-fit products. In: Paper presented at the 10th Int Research/Expert Conf, Barcelona, Spain, September

Giesel F, Mehndiratta A, Von Tengg-Kobligk H, Schaeffer A, Teh K, Hoffman E, Kauczor H, van
Beek E, Wild J (2009) Rapid prototyping raw models on the basis of high resolution computed
tomography lung data for respiratory flow dynamics. Acad Radiol 16:495–498

Griffith M, Schlienger M, Harwell L, Oliver M, Baldwin M, Ensz M, ... Smugeresky EJ (1999)
Understanding thermal behavior in the LENS process. Mater Des 20(2):107–113

Gross BC, Erkal JL, Lockwood SY, Chen C, Spence DM (2014) Evaluation of 3D printing and its
potential impact on biotechnology and the chemical sciences. ACS Publications, Washington

Guarino J, Tennyson S, McCain G, Bond L, Shea K, King H (2007) Rapid prototyping technology
for surgeries of the pediatric spine and pelvis: benefits analysis. J Pediatr Orthop 27(8):
955–960

Guo N, Leu MC (2013) Additive manufacturing: technology, applications and research needs.
Front Mech Eng 8(3):215–243

Harrysson O, Hosni Y, Nayfeh J (2007) Custom-designed orthopedic implants evaluated using
finite element analysis of patient-specific computed tomography data: femoral-component case
study. BMC Musculoskelet Disord 8:91

He J, Li D, Lu B, Wang Z, Tao Z (2006) Custom fabrication of composite tibial hemi-knee joint
combining CAD/CAE/CAM techniques. Proc Inst Mech Eng [H] 220:823–830

He J, Li D, Lu B, Wang Z, Zhang T (2006a) Custom fabrication of a composite hemi-knee joint
based on rapid prototyping. Rapid Prototyping J 12(4):198–205

He Y, Ye M, Wang C (2006b) A method in the design and fabrication of exact-fit customized
implant based on sectional medical images and rapid prototyping technology. Int J Adv Manuf
Technol 28(5):504–508

Hopkinson N, Hague R, Dickens P (2006) Rapid manufacturing: an industrial revolution for the
digital age. Wiley, Hoboken

Huang SH, Liu P, Mokasdar A, Hou L (2013) Additive manufacturing and its societal impact: a
literature review. Int J Adv Manuf Technol 1–13

Hurson C, Tansey A, O'donnchadha B, Nicholson P, Rice J, McElwain J (2007) Rapid
prototyping in the assessment, classification and preoperative planning of acetabular fractures.
Injury 38(10):1158–1162

Joshi SC, Sheikh AA (2015) 3D printing in aerospace and its long-term sustainability. Virtual Phys
Prototyping 10(4):175–185

Kalet IJ, Wu J, Lease M, Austin-Seymour MM, Brinkley JF, Rosse C (1999) Anatomical
information in radiation treatment planning. In: Paper presented at the proceedings of the
AMIA symposium

Kamrani AK, Nasr EA (2010) Engineering design and rapid prototyping. Springer, Berlin

Kettner M, Ramsthaler F, Potente S, Bockenheimer A, Schmidt PH, Schrodt M (2014) Blunt force
impact to the head using a teeball bat: systematic comparison of physical and finite element
modeling. Forensic Sci Med Pathol 10(4):513–517

Kim MS, Hansgen AR, Wink O, Quaife RA, Carroll JD (2008) Rapid prototyping. Circulation 117
(18):2388–2394

Klepáček I, Malá PZ (2012) "Bochdalek's" skull: morphology report and reconstruction of face.
Forensic Sci Med Pathol 8(4):451–459

Knox K, Kerber C, Singel S, Bailey M, Imbesi S (2005) Rapid prototyping to create vascular
replicas from CT scan data: making tools to teach, rehearse, and choose treatment strategies.
Catheter Cardiovasc Interv 65:47–53

Kobryn P, Ontko N, Perkins L, Tiley J (2006) Additive manufacturing of aerospace alloys for
aircraft structures. DTIC Document

Kruth J-P (1991) Material incress manufacturing by rapid prototyping techniques. CIRP Ann
Manuf Technol 40(2):603–614

Kruth J-P, Mercelis P, Van Vaerenbergh J, Froyen L, Rombouts M (2005) Binding mechanisms in
selective laser sintering and selective laser melting. Rapid Prototyping J 11(1):26–36

Kumar V, Dutta D (1997) An assessment of data formats for layered manufacturing. Adv Eng
Softw 28(3):151–164

Lee M, Chang C, Ku Y (2008) New layer-based imaging and rapid prototyping techniques for computer-aided design and manufacture of custom dental restoration. J Med Eng Technol 32:83–90

Lipson H (2013) Standard specification for additive manufacturing file format (AMF) version 1.1. ASTM International, 10

Liu Q, Leu MC, Schmitt SM (2006) Rapid prototyping in dentistry: technology and application. Int J Adv Manuf Technol 29(3–4):317–335

Lyons B (2014) Additive manufacturing in aerospace: Examples and research outlook. The Bridge 44(3)

Martin RL, Bowden NS, Merrill C (2014) 3D Printing in Technology and Engineering Education. Technol Eng Teach 73(8):30–35

Matias E & Rao B (2015) 3D printing: On its historical evolution and the implications for business. In: 2015 Portland International Conference on Management of Engineering and Technology (PICMET), pp 551–558

Maslowski E, Heise S (2002). President's house 3D model

Mavili ME, Canter HI, Saglam-Aydinatay B, Kamaci S, Kocadereli I (2007) Use of three-dimensional medical modeling methods for precise planning of orthognathic surgery. J Craniof Surg 18(4):740–747

Mohammed JS (2016) Applications of 3D printing technologies in oceanography. Methods Oceanogr 17:97–117

Mudge RP, Wald NR (2007) Laser engineered net shaping advances additive manufacturing and repair. Weld J-New York- 86(1):44

Munoz A, Sheng P (1995) An analytical approach for determining the environmental impact of machining processes. J Mater Process Technol 53(3–4):736–758

Nijdam TJ, Gestel RV (2010) Service experience with single crystal superalloys for high pressure turbine shrouds [Vol. Report no. (NLR-TP-2011–547, 20140118)]. National Aerospace Laboratory NLR

Noorani R (2006) Rapid prototyping: principles and applications. Wiley Incorporated

Optomec

Paiva WS, Amorim R, Bezerra DAF, Masini M (2007) Aplication of the stereolithography technique in complex spine surgery. Arq Neuropsiquiatr 65(2B):443–445

Popov I, Onuh S (2009) Reverse engineering of pelvic bone for hip joint replacement. J Med Eng Technol 33(6):454–459

Reeves P (2008) How the socioeconomic benefits of rapid manufacturing can offset technological limitations. In: Paper presented at the RAPID 2008 Conference and Exposition. Lake Buena Vista, FL

Sheng P, Munoz A (1993) Environmental issues in product design-for-manufacture. In: Paper presented at the 7th International Conference on Design for Manufacturing, Orlando

Sheng PS, Dornfeld DA, Worhach P (1995) Research: integration issues in green design and manufacturing. Manuf Rev 8(2):95–105

Singare S, Dichen L, Bingheng L, Yanpu L, Zhenyu G, Yaxiong L (2004) Design and fabrication of custom mandible titanium tray based on rapid prototyping. Med Eng Phys 26(8):671–676

Singare S, Liu Y, Li D, Lu B, Wang J, He S (2007) Individually prefabricated prosthesis for maxilla reconstruction. J Prosthodont

Singare S, Yaxiong L, Dichen L, Bingheng L, Sanhu H, Gang L (2006) Fabrication of customised maxillo-facial prosthesis using computer-aided design and rapid prototyping techniques. Rapid Prototyping J 12(4):206–213

Skelton J (2008) 3D Printers and 3D-printing technologies almanac. 2017, from http://3d-print.blogspot.com.tr/2008/02/fused-deposition-modelling.html

Sun K, Wei TS, Ahn BY, Seo JY, Dillon SJ, Lewis JA (2013) 3D printing of interdigitated Li-ion microbattery architectures. Adv Mater 25(33):4539–4543

Sun S-P, Wu C-J (2004) Using the full scale 3D solid anthropometric model in radiation oncology positioning and verification. In: Paper presented at the engineering in medicine and biology society, 2004. IEMBS'04. 26th Annual International Conference of the IEEE

Suzuki M, Ogawa Y, Kawano A, Hagiwara A, Yamaguchi H, Ono H (2004) Rapid prototyping of temporal bone for surgical training and medical education. Acta Otolaryngol 124:400–402

Symes MD, Kitson PJ, Yan J, Richmond CJ, Cooper GJ, Bowman RW, Cronin L (2012) Integrated 3D-printed reactionware for chemical synthesis and analysis. Nat Chem 4(5): 349–354

Tek P, Chiganos T, Mohammed J, Eddington D, Fall C, Ifft P, Rousche P (2008) Rapid prototyping for neuroscience and neural engineering. J Neurosci Methods 172:263–269

Truscott M, De Beer D, Vicatos G, Hosking K, Barnard L, Booysen G, Ian Campbell R (2007) Using RP to promote collaborative design of customised medical implants. Rapid Prototyping J 13(2):107–114

Tuck C, Hague R, Burns N (2006) Rapid manufacturing: impact on supply chain methodologies and practice. Int J Serv Oper Manag 3(1):1–22

Uriondo A, Esperon-Miguez M, Perinpanayagam S (2015) The present and future of additive manufacturing in the aerospace sector: A review of important aspects. Proc Inst Mech Eng [G] 229(11):2132–2147

Wagner J, Baack B, Brown G, Kelly J (2004) Rapid 3-dimensional prototyping for surgical repair of maxillofacial fractures: a technical note. J Oral Maxillofac Surg 62:898–901

Walter M, Holmström J, Tuomi H, Yrjölä H (2004) Rapid manufacturing and its impact on supply chain management. In: Paper presented at the proceedings of the logistics research network annual conference

Wang Z, Teng Y, Li D (2004) Fabrication of custom-made artificial semi-knee joint based on rapid prototyping technique: computer-assisted design and manufacturing. Zhongguo Xiu Fu Chong Jian Wai Ke Za Zhi 18:347–351

Wohlers T, Caffrey T (2013) Wohlers Report 3D printing. Wohlers Associates, Fort Collins

Wong KV, Hernandez A (2012) A review of additive manufacturing. ISRN Mechanical Engineering

Wood D (2009) Additive Layer manufacturing at Airbus-Reality check or view into the future? TCT Magazine 17(3):23–27

Woźniak K, Rzepecka-Woźniak E, Moskała A, Pohl J, Latacz K, Dybała B (2012) Weapon identification using antemortem computed tomography with virtual 3D and rapid prototype modeling—A report in a case of blunt force head injury. Forensic Sci Int 222(1):e29–e32

Zemnick C, Woodhouse SA, Gewanter RM, Raphael M, Piro JD (2007) Rapid prototyping technique for creating a radiation shield. J Prosthet Dentist 97(4):236–241

Chapter 14
Advances in Virtual Factory Research and Applications

Alperen Bal and Sule I. Satoglu

Abstract Simulation is a powerful tool to observe the performance of the manufacturing systems in variable demand and dynamic factory conditions in virtual environments. Recently, modeling and analysis of the factory objects in a 3-dimensional Simulation software emerged that is called a Virtual Factory (VF). The VF provides an integrated simulation model by representing all major aspects of a factory. It helps to consider the factory as a whole and provides decision support (Jain et al. in Winter simulation conference (WSC). IEEE, Piscataway, 2015). In other words, a VF Framework is a virtual advanced software environment that aims assisting the design and management of all physical factory entities during the all phases of the factory life-cycle (Azevedo et al. in Management and Control of Production and Logistics:320–325, 2010). In this chapter, the research projects and academic papers that focus on the Virtual Factories are reviewed. Besides, the commercial VF software are investigated, their distinguishing aspects are assessed, and their limitations are discussed. Finally, future work suggestions are presented.

Abbreviations

APS	Advanced Planning and Scheduling
CAD	Computer Aided Design
CFD	Computational Fluid Dynamics
CME	Collaborative Manufacturing Environment
DES	Discrete Event Simulation
DVF	Distributed Virtual Factory
ERP	Enterprise Resource Planning
FCM	Fatigue Cracking
IoT	The Internet of Things
KPI	Key Performance Indicator
MHS	Material Handling System
MRP	Materials Requirement Planning

A. Bal (✉) · S.I. Satoglu
Istanbul Technical University, Macka, 34367 Sisli, Istanbul, Turkey
e-mail: abal@itu.edu.tr

© Springer International Publishing Switzerland 2018
A. Ustundag and E. Cevikcan, *Industry 4.0: Managing The Digital Transformation*,
Springer Series in Advanced Manufacturing, https://doi.org/10.1007/978-3-319-57870-5_14

RFID   Radio Frequency Identification
SAM   Stress Analysis
SDM   Structural Dynamics
VF   Virtual Factory
VFF   Virtual Factory Framework
WMS   Warehouse Management System

14.1 Introduction

The Internet of things (IoT) is a paradigm that covers communication and integration of many objects that can interact with each other through the information technologies to attain certain system goals (Atzori et al. 2010). These objects are Radio-Frequency Identification (RFID) tags, sensors, actuators, mobile phones, even white goods that can all connect to the wireless networks by means of the unique addressing schemes. Emergence of the IoT paradigm has also triggered the concept of Smart Factories (Zuehlke 2010). Smart Factories intend deployment of the Cyber-Physical Systems and the information technologies so that the machines, robots, and other manufacturing resources become smart in the sense that they can communicate, collaborate, and perform more efficiently and resiliently (Lee et al. 2015). The term of Industry 4.0 implying the fourth industrial revolution is also used as a synonymous term of the Smart Factories.

In today's competitive and rapidly changing market environments, the companies must be flexible, modular, adaptable, scalable and knowledge-based and must employ information and communication technologies very well. Therefore, in order to adapt to the technological changes and demand fluctuations, before making major factory design decisions, simulation modeling of the factory may provide great insight.

Simulation is a powerful tool to observe the performance of the manufacturing systems in variable demand and dynamic factory conditions in virtual environments. Negahban and Smith (2014) reviewed the past studies that employ the simulation technique for the design and operation of the Manufacturing systems. The authors pointed out that Simulation modeling of the manufacturing systems has been extensively studied. The design papers can be categorized as general facility layout design, material handling system design (Huang et al. 2012) including the MHS design of the flexible manufacturing systems and design of the semi-conductor manufacturing systems (Jimenez et al. 2008). Besides, many papers employed simulation while designing the hybrid manufacturing systems (Durmusoglu and Satoglu 2011), cellular manufacturing systems (Chtourou et al. 2008) and flexible manufacturing systems (Mahdavi and Shirazi 2010). Planning and Scheduling of the Production and Maintenance activities, real-time simulation and control of the manufacturing systems are the other research streams in the simulation literature.

Recently, modeling and analysis of the factory objects in a 3-dimensional simulation software environment has also emerged that is called a Virtual factory.

It provides an integrated simulation model by representing all major aspects of a factory. It helps to consider the factory as a whole and provides decision support. The developed model should be capable of assessing performance at multiple grades and generate data as realistically as possible in compliance with the real factory conditions. It helps to consider the factory as a whole and provides decision support. The virtual factory concept makes it possible to focus on a single process/equipment, a department/a production line or the factory as a whole. In other words, the scales of the virtual factory extend from the factory level to the device level.

Jain et al. (2001) defined the Virtual Factory as "*An integrated simulation model of major subsystems in a factory that considers the factory as a whole and provides an advanced decision support capability.*" In addition, Azevedo et al. (2010) proposed the *Virtual Factory Framework* as a virtual advanced software environment that aims assisting the design and management of all physical factory entities including all products, manufacturing resources, and even the network of companies during the all phases of the factory life-cycle. The authors suggested that the factory life cycle management including *factory planning and optimization* should be performed in virtual environment, by considering the "*factory as a product*". Later, the detailed Digital Factory Templates of this study are explained by Azevedo and Almeida (2011). Upton and McAfee (1996) proposed that a virtual factory provides an environment for several partners of a product or process to work in coordination by sharing information, electronically. The increased grade of accuracy of facility's sub-system models provide an opportunity use of the Virtual factory for supporting decisions such as bench marking of alternative business processes and selection of communication network architects (Jain et al. 2001).

Although the terms of Digital Factory and the Virtual Factory sound similar to each other, they do not mean the same. The Digital Factory is a static image of the current state of the factory that is demonstrated by using the digital manufacturing and modeling tools and techniques. However, the Virtual Factory is a "projection into the future" of factory by means of 3-D Virtualization and Simulation techniques in software environment (Constantinescu 2011). In other words, the virtual factory term refers to the *simulation* of all production resources including the human operators in an integrated and most realistic way in a software environment, to assess different future alternatives of the factory layout, production line, cell or process designs.

A virtual factory may provide some opportunities to test the different cases of a real facility in the following cases:

- The manufacturing systems are too expensive to construct or risky to do live tests.
- A design change within a large or complex system is being considered or predicting process variability of the system is important.
- When there is an incomplete data of the manufacturing system.
- When it is needed to communicate the design ideas to help the participants of the system better understand the system.

Nevertheless, simulation of a factory with existing paradigm and technologies is difficult even with the powerful computers to process. Since a factory as a whole system consists of many subsystems and modeling of all subsystems at a desired level of detail and the creation of a whole factory model makes the simulation process quite complicated. Even a single run of simulation models can take quite a long time.

According to the report issued by ARC Advisory Group Collaborative Production Management (CPM), 2007, "In the near future, modeling and simulation will represent a new way of doing business". In accordance with these developments, some commercial software has been developed so far in order to simulate the advanced *smart factories*.

The purpose of this chapter is to explain the concept of a "Virtual factory"; discuss the possible benefits that can be earned by means of the Virtual Factory applications in software environments; to explain the examples of the current software facilities developed so far by different commercial companies, and to discuss the limitations of this software.

The chapter is organized as follows: In Sect. 14.2, the State of Art including the research papers and the Virtual factory software examples are presented. Then, the limitations of the commercial software are discussed in Sect. 14.3. Finally, the Conclusion and Future Work are presented.

14.2   The State of Art

In this section, first, the academic research papers and projects dealing with the Virtual Factories are reviewed. Then, the commercial Virtual Factory software facilities and implementations are analyzed and discussed. Hence, both the academic research and the practical applications are intended to be covered.

14.2.1   Research Papers and Projects

In this section, the journal and conference publications papers that deal with the VF and the research projects that aim at developing VF infrastructures are analyzed and discussed. One of the pioneer studies belongs to Fujii et al. (2000) where a Distributed Virtual Factory (DVF) concept was proposed comprised of the integrated and precise Simulation models of the production modules including the material handling activities. Several "distributed" simulation models were linked to each other by the synchronization mechanisms (Fujii et al. 2000).

One of the research projects is an FP6 EU Project called *Digital Factory for Human-Oriented Production Systems (DiFac)* was performed to develop a collaborative manufacturing environment (CME) of the next generation Manufacturing companies, and specifically targeted the SMEs of Europe. According to Sacco et al. (Sacco et al. 2009), the Digital Factory is a static model of the current manufacturing processes, technologies, methods and tools that are used, whereas the future state is represented in the Virtual factory. By using the data obtained out of the Digital Factory, the possible future system designs are simulated within the 3-dimensional virtual reality models. According to the results obtained by the simulation models of the VF, the new processes, methods or tools are adapted within the Digital Factory. The tools used by the VF are the tools for process modeling, simulation software, virtual reality and the Augmented reality systems (Sacco et al. 2009). The authors claimed that the integrated Digital and Virtual Factory approach may provide the following benefits: People at different locations can better cooperate on the same project. By means of the virtual new product design, production times and Waste can be reduced. Collaboration among people working on the same project in different places is facilitated at different stages of a project. The decision-making process may become interactive that facilitates the innovations. The knowledge-base pertaining to the product design may be shared and accessible by different parties. Due to the human-centric nature of the development processes, workers' efficiency and safety are enhanced by training and learning on the virtual lines and machines. Besides, training for the emergency cases can be provided to the workers in virtual environments (Sacco et al. 2009).

The research issues covered in the DiFac project is explained comprehensively, in an international book (Canetta et al. 2011). One of the interesting aspects is the Digital Human modeling where human postures and motions are virtually simulated in a 3-D manner, by either task-based modeling or data-based modeling (inverse dynamic modeling) by means of the special software (Mun and Rim 2011). Hence, the joint effort, discomfort and energy consumption are intended to be minimized while optimizing the workstation design. Constantinescu (Constantinescu 2011) explained the Factory Constructor® software developed during the DiFac project. The Factory Constructor is comprised of the Web-based decision-making tool called iGDSS, GIOVE the Virtual Factory template, the Web-based eM-Plant Simulation Tool, and the Web platform iPortal for content management (Constantinescu 2011). In addition, in this DiFac project, an Augmented Reality-based Training Simulator was developed to support the workers for the Maintenance activities and emergency cases (Buckl et al. 2011).

In addition, an FP7 European Union project called "Holistic, extensible, scalable and standard Virtual Factory Framework (VFF)" that intends to improve and reinforce the European Manufacturing Industry by defining the next generation VFFs, by enhancing the design, management and reconfiguration of the existing factories (Azevedo et al. 2010). In this project, the conceptual foundations of the VFF were proposed. Basically, the data synchronization between the real factory and the Virtual factory is aimed. The VFF software is comprised of the Reference Model, Virtual Factory Manager, Functional Modules, and Knowledge Repository (Azevedo et al. 2010). According to the results of the VF Simulation runs, the real factory is designed and operated, whereas the data obtained out of the real factory including the key performance indicators (KPIs) are fed back to the VF software and the simulation model is updated accordingly. This is called the data synchronization. As an extension of this VFF project, the factory templates were proposed so as to decrease the simulation model development time (Azevedo and Almeida 2011). Azevedo and Almeida (2011) proposed that the templates are built on the Enterprise Reference Models that are the concept and the elements common to all of the factory life-cycle. These are called Generalized Enterprise Reference Architecture and Methodology (GERAM) models. The Factory Templates are comprised of the Static and Dynamic Domains. In the static domain, the manufacturing processes/equipment, the human resources, information systems, environmental issues that affect the design process are considered. In the dynamic domain, the Process Value Estimator algorithms such as Artificial Neural Networks etc. are utilized to forecast the possible behavior of the factory under the new set of operating conditions.

Later, Jain et al. (2015) proposed a VF Prototype that is comprised of the process, machine, and Manufacturing cell levels. In the process level, the turning machining process is modeled in virtual environment where the physical kinetics of the turning process is represented. The process parameters such as the feed rate and depth of cut etc. are input to the virtual model to estimate the resulting cutting forces, energy consumption, and duration of the process. This level of the VF software is coded in Java to be integrated with the Machine Level. The machine level represents the machines and it is modeled by using the agent-based software called Any Logic®. At this level of the VF software, the machine's states are modeled as idle, part setup, batch setup, machining, part ejection and batch ejection. The machine modeled as an agent decides to which state to pass according to a State transition logic. The Manufacturing Cell Level of the VF software is developed by using the discrete-event simulation feature of the Any Logic Software. The authors proposed that additional manufacturing processes must be modeled in future studies to develop the proposed VF Prototype Later, (Jain et al. 2015).

The VF simulation software that covers simulation modeling at the manufacturing process, machine, cell/line and plant level was formerly proposed by Kühn (2006), around ten years ago. The author called this concept the Digital Factory. It was proposed that the software would include resource database where the machines, cutting tools, gauges, robots, welding guns, and manufacturing templates all comprise a library. It was proposed that the predefined 3-D objects that represent the machines, conveyors, cranes, material containers, and workers can be used to quickly construct a 3-D Virtual factory layout. Hence, the factory, production lines/cells and processes can be analyzed by the discrete event simulation; the factory flow can be optimized by considering the part routings, the distances between the production resources, the current material handling equipment specifications. The robotic cells can be simulated in a 3-D environment to perform the path planning of the robot, to program the operation steps of the robot, detect any collisions or inconvenience regarding the work environment. Besides, manual work places can be virtually modeled to perform an ergonomic analysis motion-time analysis, to generate work instructions for the operators that are compatible with the ergonomics principles and work-safety issues.

Shamsuzzoha et al. (2017) proposed a smart process monitoring infrastructure to keep track of the Key Performance Indicators (KPIs) and visualization of the data collected by means of a user-interface called the dashboards, in the virtual environment. This software was developed during the FP7 EU Project called ADVENTURE (ADaptive Virtual ENterprise ManufacTURing Environment). Authors expressed that the conceptual design, and development was performed and the design was validated through this project (Shamsuzzoha et al. 2017).

An interesting application of the virtual factories concerns the simulation modeling of the energy consumption behavior of the Manufacturing facilities (Baysan et al. 2013). A current state value-stream map pertaining to the current energy consumption of the whole facility equipment and processes can be established. Based on this, the current state simulation model of the facility is developed. By means of the thorough analysis of the energy performance of the manufacturing stages, and the transportation activities for the raw material, WIP and products, some enhanced future state scenarios are developed and models of these are constructed in the Simulation software environment. Later, based on the simulated scenarios, the solutions to improve the energy consumption of the whole production facility can be assessed (Baysan et al. 2013). All simulation modeling is proposed to be performed in Arena$^®$ discrete-event simulation software. In the next sub-section, it is mentioned that the Plant Simulation$^®$ commercial software that belongs to Siemens is able to perform energy consumption simulation and Economic analysis of the factories.

The concept of the *Digital Twin* has been also mentioned in the literature associated with the virtual simulation of the physical entities (Tao et al. 2017). Here, collected physical product data is fed into the virtual product model in order to manage the product life-cycle in a better way, including the product design, manufacturing and service. Tuegel et al. (2011) expressed that a digital twin aims at exploitation of advances in digital modeling and simulation of the physical entities in three dimensional modes and high performance computing in order to observe and test the performance of a complete structural design. The Digital Twin of the tail component of an aircraft was developed, where the models for the computational fluid dynamics (CFD), the structural dynamics (SDM), the thermodynamic, the stress analysis (SAM), and the fatigue cracking (FCM) are linked to each other in a virtual software model called the Digital Twin, so as to observe the behavior of the design under varying operating conditions (Tuegel et al. 2011). However, in conventional applications, the above-mentioned models are isolated to each other, and modeled in fragmented environments that prevent analysis of the interactions among these design aspects.

14.2.2 The Virtual Factory Software

Corresponding real factory with most of its features and operations should be represented in a Virtual factory. A variety of software is used to create a virtual factory. They allow engineers to run experiments and what-if scenarios without disturbing existing production system. Besides, these software are also used during the preliminary design studies, before a plant construction. The widely-used software are explained in this section. The main features of them are presented in Table 14.1. The software are classified according to the simulation technique employed, namely discrete-event simulation, agent-based simulation, system dynamics, and continuous Simulation; visualization type (either 2-dimensional or 3-dimensional), energy consumption analysis tools, neural network algorithms, and the optimization algorithms. When the Table 14.1 is checked, it is clear that almost all of the software have 3-D Visualization feature, and most of them have 2-D aspect as well. All of them work on the basis of Discrete Event Simulation. However, AnyLogic® and Simio® can also employ the agent-based simulation technique where the agents such as the material handling vehicles, workers, machines modelled as agents can decide how to work under predetermined conditions. This is an interesting and valued aspect while modelling the Virtual Factories. Plant Simulation® software has additional features such as Energy Consumption Analysis tool that can be used to keep track of the energy effectiveness of the whole factory and to improve it. Besides, it has the Neural Network Tool that is employed to analyze the sensor data and to forecast the failure of the equipment in advance, before a failure occurs. This is a contemporary and valued feature, as well. Besides, the Genetic Algorithms are employed by Plant Simulation® where almost an optimum material flow is reached, as the Genetic Algorithm can reach (at the best case) the optimum or near optimum solutions.

AnyLogic is a single model environment that brings system dynamics, discrete event simulation and agent-based simulation together. Their motto is released as a multimethod simulation modelling environment. It provides a flexible modelling of business, economy and social systems at the desired level of detail within one modelling language and one model development environment. It offers a variety of visual tools for modelling such as state charts, process flow charts, action charts, and stock & flow diagrams. Another advantage of this software is that it can provide very attractive visualization and animation. Therefore, the models can be presented to stakeholders in an explanatory way and can be used as an educational tool in company as well. Industry-specific libraries get the edge on other software. They namely are, process modelling library, fluid library, rail library, pedestrian library, road traffic library and material handling library. Especially Pedestrian, Rail, and Road Traffic Libraries enable detailed physical-level simulation of objects' movement and interaction, which keeps it one step ahead of its competitors in this respect. Besides, it allows importing CAD drawings, images, and shaping files into the simulation models. From the viewpoint of agent-based simulation it allows to populate large-scale models of organizations' big data with agents with personalized characteristics such as consumer behavior, individual skills, schedules, performance data, or health-related profiles. GIS maps integration can also be used to model systems like supply chains, logistics networks. Like many other simulation programs, it is capable of data Interoperability and provides support in model building. In despite of extensible and customizable platform, it requires a certain

Table 14.1 The widely-used commercial Virtual factory software

Software	Discrete event simulation	System dynamics	Agent-based Simulation	Continuous simulation	2D visualization	3D visualization	Energy consumption simulation & analysis	Neural networks	Genetic algorithms
AnyLogic®	X	X	X		X	X			
Arena®	X				X	X			
FlexSim®	X					X		X	
Plant Simulation®	X					X	X	X	X
Simcad Pro®	X			X	X	X			
Simio®	X		X		X	X			
SIMUL8®	X				X				
Visual Components®	X					X			

extent of programming capability of users. Nevertheless, this adaptability of AnyLogic software gives opportunity to design custom experiments for special requisites, or use customized algorithms and optimization engines.

Arena is one of the most commonly used simulation program for discrete event simulation (DES). It uses flowchart modelling methodology including a large library of pre-defined building blocks to model processes. Drag and drop elements and structures allow building simulation models in a fast, intuitive and easy to learn manner. Arena requires no programming assistance but this can sometimes limit the user. On the other hand, SIMAN code is accessible in professional edition to intervene in. Besides, custom templates and reusable modules can be created. It has specialized library for high-speed packaging for modelling this type of production systems in a more efficient way. Another feature of Arena simulation is to handle not only discrete but also flow/continuous manufacturing modelling and simulation.

Arena is in use for 30 years and preferred by many of the large scale companies. A variety of application areas exist from Manufacturing and, healthcare to military services. Especially 3D modelling capability helps to create the digital-twin of the facility. Arena has Virtual Designer tool for animation built from a gaming engine. Cândea et al. (2012) has been realized a Virtual factory module based on Arena. They implemented virtual factory framework (VFF) to a supplier of automotive components factory. By means of this project the tools including Arena needed to create VFF have been effectively used and expedited decision making process.

FlexSim also is a cutting-edge discrete event simulation program. It offers increased 3D capabilities with improvements in ease-of-use. Flexsim allows directly model in 3D environment while many other simulation programs require post-processing. The other prominent feature of FlexSim compared with other simulation programs is its ability to optimize. By using the Experimenter tool of Flexsim, scenarios by changing the parameters levels can be created and run and the results are reported. Besides, OptQuest optimization tool of Flexsim employs tabu search, scatter search, integer programming, and neural networks algorithms to efficiently select the best scenarios according to the parameter values specified by the designer.

Plant Simulation is used to create digital model of factory so that to create digital twin of the facility. It is a discrete-event simulation tool that enables optimization of production systems and processes. Yet it is applicable not only discrete but also continuous processes. One of the most important features of Plant Simulation are automatic bottleneck detection, calculating and optimizing energy usage via energy analysis tools, using genetic algorithm and neural networks for system optimization. The models can synchronize with supply chain, production resources and business processes with open system architecture. This enables having correct analysis by virtue of different participants of manufacturing process. Visualization of the factory is available both in 2D or 3D. Using library or computer-aided-design (CAD) data models can be visualized in 3D environment. Value stream mapping library is also available as an extension. Static, paper-based value stream mapping is depicted dynamically through this extension. Main contribution of this feature is to show value-adding and non-value-adding activities.

Siemens Automation website (Siemens PLM Automation 2017) shows the 3D modelling of an Indian automobile supplier industry company Eicher Engineering Solutions (EES) production facility. They used Plant Simulation software to create digital-twin of the plant. The layout, equipment, and product flow of the plant are modelled realistically in 3D. The findings of the prepared model showed that significant improvements can be made in plant design, utilization of space, movement of materials and planned investments. Indeed, the improvements put into practice have resulted in significant savings and increased productivity. Besides, similar application is done for sister companies in India. In one project, they could increase capacity by a factor of almost five without additional land and in another project, they increased overall space utilization by 33% in addition to a number of improvements. The applications done with this simulation program shows that very efficient and effective results can be achieved.

SimCad Pro is dynamic discrete and continuous simulation software which allows modifying the model during the Simulation run, analysing the impact of constraint changes. It is also able to be integrated with live and historical data. SimCad Pro has different tools namely SimCad process simulator, SimTrack, SimData, IAnimate3D. SimTrack interact with live data systems such as Radio frequency identification system (RFID), Bar Code, Warehouse Management System (WMS), Enterprise Resource Planning (ERP), Materials Requirement Planning (MRP). Therefore, models can become capable of learning from the current environment. This makes real-time operational status of the facility visible. The items in the facility can be tracked in real time in 2D or 3D. Dynamic analysis of the system automatically reschedules the production flow. On the other hand, SimData provides time and motion study data collection. It allows users to effectively implement time studies on existing operations. IAnimate3D is a 3-D animation software that can be used to build desired images of equipment's, tools and other components as well as realistic movement of people. Compatibility with other drawing programs makes it possible to import 3D images. SimCad Process simulator dynamically simulate manufacturing environment and also automatically generates the model value stream map. Besides making spaghetti diagrams, swim lanes, heat maps, efficiency and OEE calculation strengthen users hand to obtain powerful reporting and analysis.

Simio gives simulation solutions for discrete part manufacturing. Besides discrete event simulation, it provides a relatively modern approach agent-based simulation and 3D object-based modelling environment. Simio is optimized to develop 3D models that can construct 3D model in a single step from a top-down 2D view, and then instantly switch to a 3D view of the system. Simio also offers simulation and scheduling optimization. Simio Production Scheduling Software gives a detailed schedule considering Advanced Planning and Scheduling (APS) approach. You can model the system in a perfect environment with no machine failures, constant process times, arriving material on time and it generates a detailed schedule. Once this schedule generated Simio replicates this simulation model considering variations. Therefore, a risk-based planning and scheduling is performed to obtain probability of meeting the defined targets, as well as expected, pessimistic and optimistic schedule outcomes. Simio Production Scheduling Software can integrate with external relationship data sources such as ERP or other business transaction systems. In real life applications, it is very essential to automatically get the data into the simulator for quick action. Also, changing production schedules are often required under real production conditions. Therefore, initialization of the system with real time data, fulfil instantaneous need for a schedule change.

Thiesing and Pegden (2015) gave examples of Simio Scheduling in different sectors. BAE Systems used Simio's scheduling application for generating schedules, performing risk and cost analysis, investigating potential improvements, and viewing those parameters in 3D animations. Schneider Electric integrated Simio with their real-time manufacturing execution systems. By this integration the operators have the ability to see planned schedule along with feasible alternatives when the plan is no longer applicable due to unexpected changes.

Simul8: Another object-based discrete event simulation program Simul8 enables to effectively design, launch and improve production environment of the facility. Simul8 is easy to use with drag and drop interface and more intuitive in comparison with other simulation programs. So, it does not rely as heavily on coding such as C++, VBA and Java. In virtue of Simul8's own simulation language this makes it superior to some legacy packages still using the SIMAN processor and simulation language. Looking at the application areas, it comes into sight that many world scale companies use Simul8. Also, one of the biggest advantages of this software is that it can run faster.

Visual Components provides ease of use to the user and visualize 3D model of factory layout uncomplicatedly. It also offers many webinars, courses and lessons that can help the user. One of the most important points contributed by this program is that it can provide robot programming. It allows teaching motion points, configuring tools and bases (coordinate system), configuring and connecting inputs/outputs. It has quite a flexible and adaptable design capability specific to component in the manufacturing environment. Developed simulation models provide some KPIs to assess performance of the system.

Also, many of these programs can be used to make decisions for healthcare, manufacturing, automotive, supply chain, justice and business sectors. Yet the digital twin technology spans across all industries as well as healthcare facilities. Modelling and simulation studies applied for this purpose. Thus, customer provided decision variables; KPIs and system properties can be analyzed to effectively manage the healthcare system.

In order to design and implement advanced simulation systems, great attention must be paid to real time data of processes as well as historical data. If simulation systems integrate with data driven knowledge so the behavior of the real system can be extrapolated rapidly, several optional and solution alternatives can be established.

Looking at the developments in simulation software in recent years, it is seen that they have switched to cloud technology which is changing the way people run models. This situation provides online simulation analytics not only to employee and managers of company but also to clients. Furthermore, high-performance cloud computing for complex experiments can become applicable by this way.

14.3 Limitations of the Commercial Software

Although the VF software may provide several benefits, their implementation requires special effort and resources. These are briefly discussed below.

- The Virtual Factory simulation model must be customized for each of the manufacturing facility's specific conditions, as realistically as possible. So, this customization is a labor-intensive work and requires special software expertise.
- The Virtual Factory software may be costly to develop, especially when the production facility if complicated. In these cases, the cost to incur must be justified in return to the value that can be earned by virtual simulation modeling and analysis.
- Real-time data collection is strongly needed to keep track of the status of the whole manufacturing system, the machines, robots and other equipment, in a continuous manner, and in order to perform the simulations as realistically as possible. Hence, as soon as some parameters change within the manufacturing system, this can be quickly reflected to the simulation models.
- Data analytics tools and algorithms must be also integrated with the simulation software. However, only a few of the commercial software are able to analyze the real-time data. More algorithms such as Artificial Neural Networks etc. are required to be integrated with the software.

14.4 Conclusion

In this chapter, recent advances Virtual Factories research projects and academic papers are reviewed. Besides, the main features of the commercial Virtual Factory software are presented and discussed. It can be concluded that most of the software employ discrete-event simulation for modeling. Only a few of them (AnyLogic and Simio) employ the Agent-Based simulation. Besides, only the Plant Simulation® software has the energy consumption analysis tool that is a good utility provided to the designers to improve the energy effectiveness of the factories. Moreover, only two software products, namely the Plant Simulation® and FlexSim® have the Neural Network tool for system optimization and/or equipment failure prediction. This is also a distinguishing aspect of them. Almost all of the software have strong 3-D Visualization feature. Especially, the SimCad Pro software can be easily integrated with the ERP software; it automatically generates Value-Stream Maps and Spagetti Diagrams that are lean visual tools for communicating the performance of the factory.

When the academic literature is reviewed, it is clear that many papers proposing conceptual Virtual Factory infrastructures exist. However, the case studies that exemplify the use of the Virtual Factory software are rare. So, in future studies, more empirical evidence is needed in order to validate the utility of the VF software, and the benefits obtained. Besides, Energy analysis tools should be more extensively developed and implemented within the VF software, as the energy effectiveness is a critical issue for the sustainability of the factories.

References

Atzori L, Iera A, Morabito G (2010) The internet of things: a survey. Comp Netw 54(15):2787–2805

Azevedo A, Almeida A (2011) Factory templates for digital factories framework. Robot Comp-Integr Manuf 27(4):755–771

Azevedo A, Almeida A, Bastos J et al (2010) Virtual factory framework: an innovative approach to support the planning and optimization of the next generation factories. Manage Control Prod Logistics:320–325

Baysan S, Cevikcan E, Satoglu SI (2013) Assessment of energy efficiency in lean transformation: a simulation based improvement methodology. Assessment and simulation tools for sustainable energy systems. Springer, London, pp 381–394

Buckl K, Misslinger S, Chiabra P et al (2011) Augmented reality for remote maintenance. In: Digital factory for human-oriented production systems. Springer, London:217–234

Cândea G, Cândea C, Radu C et al (2012) A practical use of the virtual factory framework. In: 14th International conference on modern information technology in the innovation process of the industrial enterprises, Budapest, Hungary

Canetta L, Redaelli C, Flores M (2011) Digital Factory for human-oriented production systems. Springer, Berlin

Chtourou H, Jerbi A, Maalej A (2008) The cellular manufacturing paradox: a critical review of simulation studies. J Manuf Tech Manage 19(5):591–606

Constantinescu C (2011) Flexible integration of VR-based tools and simulation applications for the planning and optimization of factories and manufacturing processes. In: Digital factory for human-oriented production systems. Springer, London:187–200

Durmusoglu MB, Satoglu SI (2011) Axiomatic design of hybrid manufacturing systems in erratic demand conditions. Int J Prod Res 49(17):5231–5261

Fujii S, Kaihara T, Morita H (2000) A distributed virtual factory in agile manufacturing environment. Int J Prod Res 38(17):4113–4128

Huang CJ, Chang KH, Lin JT (2012) Optimal vehicle allocation for an automated materials handling system using simulation optimization. Int J Prod Res 50(20):5734–5746

Jain S, Fong Choong N, Maung Aye K, Luo M (2001) Virtual factory: an integrated approach to manufacturing systems modeling. Int J Op Prod Manage 21(5/6):594–608

Jain S et al (2015) Towards a virtual factory prototype. In: Winter simulation conference (WSC), IEEE:2207–2218

Jimenez JA, Mackulak GT, Fowler JW (2008) Levels of capacity and material handling system modeling for factory integration decision making in semi-conductor wafer fabs. IEEE Trans Semicond Manuf 21(4):600–613

Kühn W (2006) Digital factory-simulation enhancing the product and production engineering process. In: Simulation conference 2006. Proceedings of the winter simulation conference, IEEE:1899–1906

Lee J, Bagheri B, Kao HA (2015) A cyber-physical systems architecture for industry 4.0-based manufacturing systems. Manuf Lett 3:18–23

Mahdavi I, Shirazi B (2010) A review of simulation-based intelligent decision support system architecture for the adaptive control of flexible manufacturing systems. J Artif Int 3(4):201–219

Mun JH, Rim YH (2011) Human Body Modeling. Digital factory for human-oriented production systems. Springer, London, pp 165–186

Negahban A, Smith JS (2014) Simulation for manufacturing system design and operation: literature review and analysis. J Manuf Syst 33(2):241–261

Sacco M et al (2009) DiFac: an integrated scenario for the digital factory. In: Proceedings of international technology management conference (ICE). IEEE:1–8

Shamsuzzoha A, Ferreira F, Azevedo A et al (2017) Collaborative smart process monitoring within virtual factory environment: an implementation issue. Int J Comp Integr Manuf 30(1):167–181

Siemens PLM Automation, (2017). http://www.plm.automation.siemens.com/en_us/about_us/success/case_study.cfm?Component=65345&ComponentTemplate=1481#lightview%26url=/en_us/Images/EicherEngineeringSolutions_Large_tcm1023-65343.jpg%26title=Eicher_large%26description=EicherEngineeringSolutions%26width=800%26height=596.0%26docType=jpg. Accessed on 4 May 2017

Tao F, Cheng J, Qi Q, Zhang M et al (2017) Digital twin-driven product design, manufacturing and service with big data. Int J Adv Manuf Tech. doi:10.1007/s00170-017-0233-1

Thiesing R, Pegden C (2015) Simio applications in scheduling. In: Proceedings of the winter simulation conference

Tuegel EJ, Ingraffea AR, Eason TG et al (2011) Reengineering aircraft structural life prediction using a digital twin. Int J of Aerosp Eng. doi:10.1155/2011/154798

Upton D M, McAfee A P (1996) The real virtual factory, Harvard Bus Rev, July–August

Zuehlke D (2010) SmartFactory-towards a factory-of-things. Ann Rev Control 34(1):129–138

Chapter 15
Digital Traceability Through Production Value Chain

**Aysenur Budak, Alp Ustundag, Mehmet Serdar Kilinc
and Emre Cevikcan**

Abstract The value of Digital Traceability technologies are critical for the production systems, due to the high levels of complicated manufacturing processes and logistics operations. In this chapter, digital traceability technologies are discussed to analyze manufacturing processes, such as: tracking work-in-progress, tracking inventories, counting stock, receiving, picking, and shipping of semi finished products. To do this, an architectural framework of the technologies in production industry was designed. A roadmap for digital traceability design, configuration and deployment is presented in order to design an integrated approach as pointed out in the literature and in former projects. It can be clearly pointed out that, by using digital traceability technologies in production value chain, redundant inventory and production costs, redundant labour costs caused by inefficient production activities, inaccuracies of records, incorrect order deliveries and penalty costs incurred by customers could be significantly reduced, which provides invaluable advantages in the real life competition of the productions systems.

15.1 Introduction

Information technology has changed considerably that companies resisting to use of it in global competitive environment may easily lose their power. There are many new technologies that surprise users by providing vital support as well as enabling

A. Budak (✉) · A. Ustundag · E. Cevikcan
Industrial Engineering Department, Management Faculty,
Istanbul Technical University, Macka, Istanbul, Turkey
e-mail: abudak@itu.edu.tr

A. Ustundag
e-mail: ustundaga@itu.edu.tr

E. Cevikcan
e-mail: cevikcan@itu.edu.tr

M.S. Kilinc
Industrial Engineering Department, Oregon State University, Corvallis, USA

© Springer International Publishing Switzerland 2018   251
A. Ustundag and E. Cevikcan, *Industry 4.0: Managing The Digital Transformation*,
Springer Series in Advanced Manufacturing, https://doi.org/10.1007/978-3-319-57870-5_15

different and better processes. Digital Traceability systems are the significant technologies that provides remarkable benefits such as increased traceability, accuracy, security, efficiency, real time controllability and reduced labor cost.

Digital Traceability applications have been attracting R&D improvements and are practical in real world cases. In the simplest terms, Digital Traceability based systems have been applied in many different areas but have been especially utilized in logistics, supply chain management systems and production for identification, tracing and tracking (Lim et al. 2013). These conveniences provide monitoring of the system with more precise and real time data in different fields. Digital Traceability adopting systems provide increasing capacity and labour productivity with reduced operational mistakes (Ngai 2008a, b).

By utilizing Digital Traceability implementation in production systems, Digital Traceability based applications could be able to be adopted practically and efficiently for real time tracking of the malfunctions and problems that occur in production processes. Additionally, incorrectness in records and order deliveries that results in complexity in planning activities and penalty costs incurred by customers is significantly prevented, and this circumstance ensures great advantages for global competition.

Specifically, difficulties related to tracing, tracking and monitoring WIP goods and errors that occur in these processes are some of the major problems of the production systems. A common conventional method used for tracking and tracing an WIP product in the industry is the barcode system or, more primitively, manual counting. During the production process, errors caused by either barcode systems or manual counting lead to unreliable and inconsistent records whose counts are very different from the actual level of the WIP inventory. In general, the difficulty of monitoring the complicated processes and the inconsistency of records cause high labour costs and low productivity. Moreover, the inability to prevent the theft, unexplained product loss and incorrect product delivery are indirect causes of an insufficient tracking and tracing ability. As a result, dissatisfaction with the system's efficiency and with the quality of processes is a feature of this industry. Digital Traceability technologies has been utilized to solve similar problems in many different industries.

By the establishment of a Digital Traceability tracking systems early detection and control of malfunctions and problems are enabled, which provides significant advantages to production systems in today's competitive market conditions.

15.2 Digital Traceability Technologies

Automatic identification (auto-ID) technologies have the capability to track and trace objects, products, assets, and individuals through-out the value chain by capturing and transferring data with limited or no human intervention. They can provide the essential information (e.g., location, vibration, temperature, humidity, arriving time, speed, and vehicle status) in an automated and timely manner which allows early decisions and eliminates error-prone manual activities (Basole and Nowak In press; Shamsuzzoha et al. 2013). While there are several auto-ID technologies involved in automated tracking, the most commonly used technologies at various points in the production value chain are barcodes, Radio Frequency Identification (RFID), Real-time Locating Systems (RTLS), and Global positioning systems (GPS). Such technologies have different features and they can be implemented in several ways for various purposes.

1. *Barcode*: Barcodes are printed digital codes which consist of black parallel lines (bars) or small modules (squares) arranged on a white background. A barcode can be read by a barcode reader which uses a laser beam as well as a mobile device with a barcode reader software installed. While traditional barcodes with parallel lines have a low data capacity, 2D barcodes with modules (e.g., quick response (QR) code) can store up to 7089 characters. Barcode technology is inexpensive and the most commonly used auto-ID technology in the world. However, the line-of sight reading requirement, poor data security, deterioration due to dirt and bending in difficult environments, and read-only capability are the main weaknesses of the technology.

2. *Radio Frequency*: *Identification (RFID)*: An RFID system, which uses radio waves for data transfer, typically consists of the following elements:

- A tag with a chip that can store the unique identification number and information of the item to which it is attached.
- A reader with antennas to simultaneously communicate with multiple tags within the range of the antennas. The communication can be either unidirectional or bidirectional.
- Application software with a database that processes and integrates the collected information.

Unlike barcodes, which must be oriented towards the scanner for reading, RFID tags can be read without line-of sight requirements as long as they are within range of a reader. Hence, tagged objects can be easily tracked in the supply chain without human involvement. Also, RFID tags can hold greater amounts of data which can be transmitted fast, rewritten many times, and encrypted. Other advantages of RFID tags are that they are reusable and more durable to temperature and other environmental factors. Along with such benefits, the high investment cost is the main drawback of RFID in many applications. Other concerns regarding material types (e.g., metal and liquid), inference, bulk reading reliability, and lacking standardization also limit the applications.

There are different kinds of RFID tags available with different configurations, cost implications, and performance trade-offs. However, RFID systems can be broadly categorized as passive, semi-passive, or active type.

A passive tag, the simplest form of RFID tag, does not have an internal power source. This type of tag converts radio frequency energy coming from the reader antenna into electrical energy to send back a signal to the reader. Passive RFID systems can operate in the low frequency (LF), high frequency (HF), or ultra-high frequency (UHF) radio bands. They typically have short read ranges (a few centimeters for LF and up to 12 m for UHF) and can hold a little more information than barcodes. However, they are cheaper, lighter, smaller, and easier to print than other RFID types. A passive RFID solution can be useful for many applications such as identification and tracking of individuals, animals, and products within the supply chain; inventory management in the retail industry; identification of equipment and assets; and authentication of pharmaceuticals. Depending on the application, passive RFID tags can be designed in different forms. They can be mounted on a substrate, sandwiched between an adhesive layer and a paper label, embedded in a plastic card, specially packed to resist several environmental factors. Because they do not have their own energy source, sensors cannot be attached on passive RFID tags.

A semi-passive RFID tag is a battery assisted passive tag that contains an integrated power source to supply energy to the chip but relies on the reader's energy to send a signal. Semi-passive tags can have longer read ranges and some of them can support sensors, but they are more expensive and bigger than passive tags.

An active RFID tag contains its own transmitter and a power source which can be a battery, or a solar panel to draw energy form the light. This type of tag can broadcast a signal independently to a reader. Active RFID systems usually operate in the UHF radio bands and have substantially larger ranges (up to 100 m). They can transmit their location and other important information. The cost of an active tag can vary from a few dollars to a hundred dollars, depending on the reading range, battery life, durability, and other capabilities required. Readers also are more expensive than the ones used for passive systems. Active tags are generally used on expensive and large assets (e.g., trucks, containers, rail cars, animals, medical equipment). The on-board power source can also support sensor operations which greatly improves the utility of the RFID systems, especially for cold chain logistics of temperature sensitive products like fruits, fish, drugs, and vaccines (Ruiz García and Lunadei 2010).

Transponders and beacons are the two main types of active tags. Transponders are powered on when they receive a signal from a reader. For example, they are used for vehicle identification in secured gates. When a car with a transponder approaches a gate, the transponder on the car wakes up with the signal of the reader at the gate and then sends its unique ID to the reader. Transponders conserve battery life as they send signal only when they are within the range of a reader. Beacons, on the other hand, continuously broadcast signals at pre-set intervals such as every few seconds or once a day (Poslad 2009).

3. *Real-time Locating Systems (RTLS)*: Real-time Locating Systems (RTLS) provide the location of objects or people in real-time by continuously tracking them within a building or other closed area. The information regarding the tracked object or person can include the location, speed, temperature, and other specified information. Examples of RTLS include tracking patients in a hospital, tracking automobiles through an assembly line, or locating lift trucks in a warehouse. Current RTLS are based on wireless technologies such as RFID, infrared, ultrasound, and wireless sensor networks (Shamsuzzoha et al. 2013).

Among different RFID types, beacons (active systems) are often used in RTLS. The location of the tagged object can be identified by reading the signal of the tag by at least three antennas. Passive systems can provide low-cost RTLS solutions. However, the data in the passive tags can be read only at specific locations within the monitored area.

Infrared technology uses non-visible lights to detect whether an object or a person is in its location or not. Infrared rays cannot pass through the walls and they require line-of-sight transmission. To improve accuracy several infrared receivers can be installed in the building monitored. Infrared light is easily transmitted and received, and the technology offers high accuracy with almost zero false reading (Budak and Ustundag 2015).

Ultrasound, which has a high frequency beyond the hearing level, can be used in RTLS. Tags attached to the surface of objects or people transmit a unique ultrasound signal to communicate their locations to the microphones places around the room being monitored. Similar to infrared waves, ultrasound signals provide accurate results with careful positioning of several receivers (microphones), and they are contained by walls.

Wireless sensor networks (WSN) consist of small and low-powered autonomous computing nodes which are spatially distributed for monitoring and recording conditions at diverse locations. Every computing node is equipped with a transducer, microcomputer, radio frequency transceiver, power source; and a sensor which detects light, temperature, humidity, sound, pressure, vibration, pollutant levels, or other physical conditions. The collected data are transmitted among sensor nodes by hopping, and finally reach the gateway for connection to the internet. WSN can be designed in different network topologies including star, where each node connects directly to a gateway; cluster tree, where each node connects to a node in higher tree and then to a gateway; mesh, where nodes can connect to multiple nodes and the message goes through the most reliable path available; and ring, where every node is connected to two neighbors (Dwivedi and Vyas 2011; Sharma et al. 2013).

4. *Global positioning systems (GPS)*: GPS allow items to be remotely tracked anywhere on Earth at real-time using satellite navigation system. GPS are appropriate to track vehicles and shipping containers in logistics. A GPS receiver can determine its location in latitude and longitude by communicating to at least three satellites and calculating the distance to each (Shamsuzzoha et al. 2013). The location information is then transferred to a server through a wireless page/voice network (Johnson 2008). A GPS receiver can operate worldwide for several years.

15.2.1 Architectural Framework

In this section, the architectural framework of the Digital Traceability system will be explained. This framework has five cross sectional layers: the physical layer, data capturing front end, data capturing layer, processing modules and application layer. The layers are shown in Fig. 15.1, and detailed explanations are shown below.

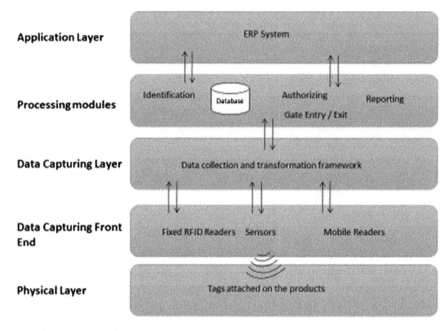

Fig. 15.1 Architectural framework of the Digital Traceability systems

1. *Physical Layer* : The first layer includes the tags that contain information about the products.
2. *Data Capturing Front End* : The second layer includes readers, sensors which are used for data collection. As the work in process is received or sent from the warehouse or related production unit, the real time information is captured by the data capturing front end layer. The reader has system software that provides interaction between the reader and tag.
3. *Data Capturing Layer*: The third layer is a kind of software that acts as a bridge between the data capturing front end units and the processing modules. The basic functions of this layer are,

- Converting Digital Traceability system information to business information
- Distributing Digital Traceability system information to business processes
- Filtering information
- Triggering business process from Digital Traceability system process and vice versa
- Managing Digital Traceability system devices
- Consolidating Digital Traceability system information

These functions provide clean, accurate and useful information from the huge amount of data captured by the readers.

4. *Processing Modules*: The forth layer is the processing modules, which request Digital Traceability system and other unit events from the data capturing layer and generate events and reports. This includes data, such as historical, time and location data. Some event and reports of this layer are picking and receiving activities, authorization, a work in process record and inventory counting. The events and reports are stored in a relational database system and shared with the application layer.

5. *Application Layer*: The last layer is the application layer where the core business processes are fulfilled. The application software at this layer includes Enterprise Resource Planning (ERP) and WEB applications and supports the business processes. The required application interfaces are all located at this level. These interfaces allow the user to utilize the applications, such as by entering the related data into the computer for gathering the inventory detail from the system. ERP is the main application of this level, and some important existing user interfaces allow the user the functions as listed below:

- Associating the product information
- Updating the record according to the received and leaving products from the warehouse and related production units
- Querying the stock level of products from the system
- Counting the physical inventory
- Querying the products being taken to/from the system and related production units

15.3 Applications

Industry 4.0 makes production enviroment more intelligent, flexible, and dynamic by equipping manufacturing with sensors, actors, and autonomous systems [8]. Therefore, machines and equipment will achieve high levels of self-optimization and automation. Self-optimization and automation level will be increase and production capacity achieve more complex and qualified standars.

Thus, the main aim of industry 4.0 is creating intelligent factories and smart manufacturing (Roblek 2016). Agent paradigm is recognized as one of the effective tools for smart manufacturing. Adeyeri et al. (2015), identify the trends of the usage of agents and multi-agents in manufacturers' resource planning and offer a framework.

Industry 4.0 creates value added integration to manufacturing processes in two ways like horizontal and vertical. Horizontal part creates modules from the material flow to the logistics of product life cycle, on the other hand vertical procedure integrates product, equipment, and human needs with different aggregation levels Shafiq (2015).

Oses et al. (2016) developed a model for an injection machine to estimate the adjusted baseline with lower risks and uncertainties in measuring and verifying energy conversation. Shafiq et al. 2015 created a virtual manufacturing system and considered three levels: virtual engineering objects, virtual engineering processes, and virtual engineering factories. The integrated mechanism of the three levels will be helpful for building the structure of Industry 4.0 and for achieving a higher level of intelligent machines, industrial automation, and advanced semantic analytics.

Thames and Schaefer (2016), consider Software-Defined Cloud Manufacturing system which is based on leveraging abstraction between manufacturing hardware and cloud-based applications, services, and platforms. Scheuermann, Verclas & Bruegge (2015) proposed an agile factory prototype which transfers agile software engineering techniques to the domain of manufacturing; they also propose a framework that depicts the impact and feasibility of customer changes during assembly-time. Paelke (2014) introduced an augmented reality system that supports human workers in a rapidly changing production environment. S. Wang et al. (2016) analyzed the relationship between lean production and intelligent manufacturing, and proposed a lean intelligent production system (LIPS) to improve production quality and efficiency and to reduce costs, in Industry 4.0.

Digital Traceability system and related applications have been attracting R&D improvements and are practical in real world cases. In the simplest terms, Digital Traceability based systems have been applied in many different areas but have been especially utilized in production systems for identification, tracing and tracking (Lim et al. 2013). These conveniences provide monitoring of the system with more precise and real time data in different fields. Digital Traceability adopting systems provide increasing capacity and labour productivity with reduced operational mistakes (Ngai 2008).Two of the most cited review studies related to Digital Traceability system implementation are the studies belonging to Ngai et al. (2012) and Lim et al. (2013) who scanned papers from 1995–2005 and 1995–2010 respectively. From the reflections of these two studies, analysed papers mostly occurred in the logistics and SCM field. In addition to this, Ngai et al. (2012) mentioned that RFID was implemented for animal detection, aviation, building management, construction, enterprise feedback control, fabric and clothing, health, food safety warranties, library services, museums and retailing. Gandino et al. (2007) applied RFID technology in the agri-food chain for a better tracing of goods. Different from these studies, Badpa et al. (2013) used RFID technology in disaster management, and they indicated the lack of a precise and efficient identification system that could speed up search and rescue activities. Another example for RFID applications can be seen in Kim and Kim's study (2012). They suggested an RFID based location sensing system for safety management in the steel industry when surrounding environmental conditions affect RFID signals. Finally, Zhu et al. (2012) described managerial issues in RFID applications, such as personal ID and access control, supply chain and inventory tracking, motorway tolls, theft control, production control and asset management. As an important application phase of RFID implementation, different applications for RFID based information system infrastructures were discussed in the literature. For instance, Ngai et al. (2008a, b) described the design and development of an RFID-based sushi management system in a sushi restaurant to support operational efficiency, and they demonstrated an architectural framework of the RFID based information system. Additionally, Tan and Chang (2010), represented an RFID based mobile e-restaurant system by the use of a wireless local area network (WLAN) and database technologies in order to enhance quick responses.

Wang et al. (2007) presented an example of RFID tracing systems to improve the efficiency of tracking tires both in warehouse management and production processes. Amaral et al. (2011) proposed a mobile software framework for RFID based applications in order to facilitate RFID integration to business operations. Hinkka and Tätilä (2013) investigated materials handling and tracing perception in the construction supply chain by applying a survey and face to face interviews.

RFID implementation practices in the manufacturing mostly comprised the sub processes in supply chain elements as in Azevedo et al. (2014). Azevedo et al. (2014) gave diversified applications of RFID in the textile supply chain. According to this study, RFID is used to track garments and in handling processes and tracking work in progress. Guo et al. (2015) developed a radio frequency identification (RFID)-based intelligent decision support system architecture to handle production monitoring and scheduling in a distributed manufacturing environment. A pilot implementation was applied in distributed clothing manufacturing. Ngai et al. (2012) analysed a case study, which was an RFID-based manufacturing process management system in a garment factory in China. Lee et al. (2013) developed an RFID-based Resource Allocation System (RFID-RAS) integrating RFID technology and a fuzzy logic concept and applied this system to Hong-Kong based garment manufacturing. Velandia et al. (2016) investigated the feasibility of implementing an RFID system for the manufacturing and assembly of crankshafts. By using RFID, the manufacturing, assembly and service data was captured through RFID tags and stored on a local server. This also could be integrated with higher-level business applications. Chen et al. (2013) proposed the integration of lean production and radio frequency identification (RFID) technology to improve the efficiency and effectiveness of warehouse management. It is stated that integration of RFID to lean, the total operation time can be saved by 87%. Moreover, the benefit of using RFID in the warehouse management is realized and promoted. Ramadan et al. (2016) created innovative real-time manufacturing cost tracking system (RT-MCT) which integrates the concepts of lean manufacturing and RFID. Zadeh and Kasiri (2016) developed production-inventory model and two scenarios were analysed on which technology should be deployed to optimize replenishment decisions one of them was barcode technology and the other was radio frequency identification (RFID) technology. Wang and Wong (2010) used RFID reader as detecting sensor, the aim of study was developed RFID-assisted object tracking system especially in flexible manufacturing assembly line.

Shin and Eksioglu (2015), analyzed labor productivity of RFID adopted retailers. As a result, there is an association between RFID technology and adopted retailer's labor productivity. The regression analysis is used and it is observed that RFID retailers have a higher labor to gross income elasticity than their non-RFID counterpart. Wang et al. (2016), put forward that most of previous studies about warehouse management system ignore the inaccuracy of the inventory and focus on this gap. Rekik, Y. (2006) also mentioned about inaccuracy of the inventory and present the reasons of errors for the inaccuracy; transaction errors, misplacement errors, damage and spoilage, theft and supply errors and it is indicated that RFID technology eliminate the inventory record errors. For instance, Poon et al. (2009) described a RFID case based logistics system in a dynamic warehouse that enables resource location determination, as well as automatic assignment of them. Also, Ren et al. (2011) indicated the lack of accurate, active and embedded information flow in construction sector for material planning and monitoring inventory. With integrating RFID technology, real time material planning and dynamic ordering process could be applied in current site working system. Liu et al. (2010), used RFID for bobbin quality controlling tracing. The information of bobbins are identified and recorded automatically in this system via non-touching dual-directional data communication based on RFID technique. As a result it is understood that RFID can use to a better control over the movement, storage and flow of materials within the warehouse by maximizing the efficiency of the receipt and shipment of goods.

15.4 Project Management in Digital Traceability

To design a project management in digital traceability; design, implementation and deployment stages of an digital traceability system for any application area should be investigated. Therefore, in this section, an integrated roadmap covering the preparation and planning stage of an traceability project, the implementation and deployment operations and system maintenance and recruitment processes is proposed according to the review of the existing literature.On the other hand methodology given in the (OTA Training 2006) does not contain the whole financial and organizational aspects. Thus, in this study, an integrated project management roadmap is designed which also includes financial and organizational elements. In financial part, cost and benefit factors, evaluation methods of alternatives in economically, sensitivity and risk analysis are considered Ustundag and Tanyas (2009). Organizational part contains determination of project goals and objectives, establishment of the project team and determination of project boundary. Table 15.1 shows the stages of each phase of the project management. The methodology enables the early-detection and prevention of problems that could occur throughout the project implementation stages or during the completed stage of digital traceability based system usage.

1. *Preparation and Planning*

   In the first phase of the roadmap, generally, a company should gather additional information about digital traceability projects, build up its digital traceability project team and determine project resources, such as the project budget and required work force. Consequently, pre-feasibility studies should be done, and supplementary contributions to the project should be investigated. In the following phase, the

Table 15.1 Proposed roadmap: Digital Traceability system design, integration and deployment stages (OTA 2006), (Ustundag and Tanyas 2009)

Digital Traceability System Design Integration and Deployment	
Stages	**Concept**

Preparation & Planning	• Gain knowledge on technologies
	• Identification of problems
	• Determination of requirements
	• Determination of project goals and objectives
	• Establishment of the project team
	• Determination of the project boundary (scope-budget)
	• Provide necessary support to the project
	• Pre-feasibility study
	  - Technical
	  - Economic
	  - Organizational
	• Preparation of the project plan and assignment of resources to the project

Development & Deployment	• *As-Is State Analysis* - Data Collection (organizational charts, process diagrams, performance criteria determination, collection of data related to machinery and labour, equipment and materials research) - Process and environmental analysis • *Design* - Solution proposals and new process development - Determination of working frequency, - Determination of tag alternatives, - Determination of hardware requirements, - Determination of pre-test scenarios, preparing machinery and performing pre-tests (tag endurance, reading performance etc.) - Specifying tag and antenna settlement, - Assigning convenient hardware (reader, antenna and tag etc.) and software - Identifying modifications that should be done to processes, - Specifying reading control points, - Detailed feasibility study • Economic feasibility -Cost Factors -Benefit Factors -Economic Evaluation Methods -Sensitivity Analysis -Risk Analysis • Technical feasibility (pre-tests related to hardware and tags) • *Verification* - Development of the software, - Pilot study (testing the solution) • Determination of pilot scheme process • Hardware procurement • System integration • Trial runs • Tests • System recruitment studies

(continued)

Table 15.1 (continued)

| Digital Traceability System Design Integration and Deployment | |
Stages	Concept
	- Solution implementation and dissemination
Maintenance and Recruitment	• Observing end user and customer experiences • Observing processes and problem determination related to the new process and solving the problems • Comparison of the planned and actual results and determination of the fields that need developments

second phase, as-is state analysis and design and deployment are the three major concepts that directly affect the overall flow of the digital traceability implementation project. Data gathering from operations and process-environment analysis are essential for current situation analysis. Besides this, determination of technology selection criteria, tag alternatives and tag and reader placement and identification of reader-additional hardware are other substantial points that play a key role for the digital traceability based system's performance. Additionally, pre-test scenarios and the specification of reading control points are other steps of the reading performance evaluation.

2. *Development and Deployment*

In the deployment stage, digital traceability based system software that controls the information flow of system tags to computers and provides end user-system interaction and pilot application is used. In this way, the digital traceability implementation process will be successfully adopted by the whole system. First of all, a pilot study or process should be carefully defined for the following procedures and to procure essential hardware and conduct digital traceability project experimental tests. After all pilot tests are applied, system recruitment should be done in order to take corrective actions and disseminate the proposed implemented system. Thereafter, the economic and technical feasibility of investment is required for determining the final decision. Moreover risk analysis is performed for investment. For the risk analysis, benefits are select and can be used in simulation as variables. Simulation models can be used to forecast investment outcomes to understand the possibilities surrounding the investment exposures and risk. In addition, running a simulation model creates a probability distribution or risk assessment for a given investment. Also it selects variable values at random and simulate model and give a known range of values but uncertain value for any particular event (Emmet and Goldman 2004).

3. *Maintenance and Recruitment*

Finally, digital traceability based system reflections should be carefully observed and evaluated by analysing end user interpretations. Consequently, problems occurring in the system should be corrected with the contribution and collaboration of both end users and the individual responsible for system implementation.

15.5 Conclusion

In this chapter, a roadmap for digital traceability design, Automatic identification (auto-ID) technologies and its applications in manufacturing system are presented in order to fill a gap pointed out by literature reviews and former projects.

The goals of Industry 4.0 are to achieve a higher level of operational efficiency and productivity, as well as a higher level of automatization. Within the application of the digital traceability based project implemented in the manufacturing industry, redundant inventory and production costs, redundant labour costs (caused by inefficient warehouse and production operations and non-value added activities, such as an inadequate control system causing a loss of products and an insufficient traceability and visibility of goods), inefficient manual processes in work processes, inaccuracies of records and difficulties in planning activities (due to inaccurate data and a lack of real time data records) will be decreased.

The detailed information of the most commonly used technologies at various points in the production value chain which are barcodes, Radio Frequency Identification (RFID), Real-time Locating Systems (RTLS), and Global positioning systems (GPS) are analyzed. An architectural framework is discussed and physical layers are considered.

In this chapter, it is clearly demonstrated that digital traceability technologies can be implemented in the production industry, and the validation of the proposed implementation roadmap is provided by the successfully designed.

Finally, it can be a reference model for understanding efficiency improvement in manufacturing system using better management philosophy and modern technology.

References

Adeyeri MK, Mpofu K, Olukorede TA (2015, March) Integration of agent technology into manufacturing enterprise: a review and platform for industry 4.0. In Industrial engineering and operations management (IEOM), 2015 international conference: 1–10. IEEE

Amaral LA, Hessel F (2011) Cooperative CEP-based RFID framework: a notification approach for sharing complex business events among organizations. In: Proceedings of IEEE fifth international conference of radio frequency identification (RFID '11)

Azevedo SG. S, Carvalho HH, Cruz-Machado V (2014) A cross case analysis of RFID deployment fast fashion supply chain. Adv Int Syst Comput 45–57

Badpa A, Yavar B, Shakiba M, Mandeep JS (2013) Effects of knowledge management system in disaster management through RFID technology realization

Budak A, Ustundag A (2015) Fuzzy decision making model for selection of real time location systems. Appl Soft Comput 36:177–184. doi:10.1016/j.asoc.2015.05.057

Chen JC, Cheng CH, Huang PB, Wang KJ, Huang CJ, Ting TC (2013) Warehouse management with lean and RFID application: a case study. Int J Adv Manuf Technol 69(1–4):531–542

Dwivedi AK, Vyas OP (2011) Wireless sensor network: at a glance. In: Lin J-C (ed) Recent advances in wireless communications and networks. InTech: 299–326

Emmet HL, Goldman LI (2004) Identification of logical errors through Monte-Carlo simulation. In: Proceedings of EuSpRIG, Conference Risk Reduction in End User Computing

Gandino F, Montrucchio B, Rebaudengo M, Sanchez ER (2007) Analysis of an RFID-based information system for tracking and tracing in an agri-food chain. In: Ozok AF, Ustundag A (eds), Proceedings of the 1st RFID Eurasia Conference: 143–148

Guo ZX, Ngai EWT, Yang C, Liang X (2015) An RFID-based intelligent decision support system architecture for production monitoring and scheduling in a distributed manufacturing environment. Int J Prod Econ 159:16–28

Hinkka V (2013) Implementation of RFID tracking across the entire supply chain. Aalto University publication series, Doctoral dissertation

Johnson ME (2008) Ubiquitous communication: tracking technologies within the supply chain. In: Taylor GD (ed) Logistics engineering handbook. (CRC Press)

Kim M, Kim K (2012) Automated RFID-based identification system for steel coils. Prog Electromag Res 131: 1–17

Lee CKH, Choy KL, Ho GT, Law KMY (2013) An RFID-based resource allocation system for garment manufacturing. Expert Syst Appl 40(2):784–799

Lim MK, Bahr W, Leung S (2013) RFID in the warehouse: a literature analysis (1995–2010) of its applications, benefits, challenges and future trends. Int J Prod Econ 145:409–430

Liu JH, Gao WD, Wang HB, Jiang HX, Li ZX (2010) Development of bobbin tracing system based on RFID technology. J Text Inst 101(10):925–930

Ngai EWT, Suk FFC, Lo SYY (2008a) Development of an RFID-based sushi management system: the case of a conveyor-belt sushi restaurant. International. Int J Prod Econ 112(2): 630–645

Ngai E, Moon KK, Riggins FJ, Yi, CY (2008b) RFID research: an academic literature review (1995–2005) and future research directions. Int. J. Product Econom 112: 510–520

Ngai EWT, Chau DCK, Poon JKL, Chan AYM, Chan BCM, Wu WWS (2012) Implementing an RFID-based manufacturing process management system: lessons learned and success factors. J Eng Tech Manag 29(1):112–130

Oses N, Legarretaetxebarria A, Quartulli M, García I, Serrano M (2016) Uncertainty reduction in measuring and verification of energy savings by statistical learning in manufacturing environments. Int J Interact Des Manuf (IJIDeM): 1–9

Training OTA (2006) RFID + Exam cram. A system's approach to RFID Implementation, Pearson

Paelke V (2014, September) Augmented reality in the smart factory: supporting workers in an industry 4.0. Environment. In Proceedings of the 2014 IEEE emerging technology and factory automation (ETFA): 1–4. IEEE

Poon TC, Choy KL, Chow KH, Lau HCW, Chan FTS, Ho KC (2009) A RFID case-based logistics resource management system for managing order-picking operations in warehouse's, expert systems with applications 36 :8277–8301

Poslad S (2009) Ubiquitous Computing: smart devices. In Environments and interactions. John Wiley & Sons, Chippenham

Ramadan M, Al-Maimani H, Noche B (2016) RFID-enabled smart real-time manufacturing cost tracking system. Int J Adv Manuf Technol: 1–17

Ren Z, Anumba CJ, Tah J (2011) RFID-facilitated construction materials management (RFIDCMM)—a case study of water-supply project. Adv Eng Inform 25(2):198–207

Rekik Y, Sahin E, Dallery Y (2006) Analysis of the impact of the RFID technology on reducing product misplacement errors at retail stores. Int J Prod Econom

Roblek V, Meško M, Krapež A (2016) A complex view of industry 4.0. SAGE Open, 6(2), 2158244016653987

Ruiz García L, Lunadei L (2010) Monitoring cold chain logistics by means of RFID. In: Turcu C (ed) Sustainable radio frequency identification solutions. InTech, Vukovar, Croatia

Sanders A, Elangeswaran C, Wulfsberg J (2016) Industry 4.0 implies lean manufacturing: research activities in industry 4.0 function as enablers for lean manufacturing. J Ind Eng Manage, 9(3)

Scheuermann C, Verclas S, Bruegge B (2015, August) Agile factory-an example of an Industry 4.0 manufacturing process. In cyber-physical systems, networks, and applications (CPSNA), 2015 IEEE 3rd international conference :43–47. IEEE

Shafiq SI, Sanin C, Toro C, Szczerbicki E (2015) Virtual engineering object (VEO): toward experience-based design and manufacturing for Industry 4.0. Cyber Syst 46(1–2): 35–50

Shamsuzzoha AHM, Ehrs M, Addo-Tenkorang R et al (2013) Performance evaluation of tracking and tracing for logistics operations. Int J Shipping Trans Log 5:31–54

Sharma D, Verma S, Sharma K (2013) Network topologies in wireless sensor networks: a review. Int J Electron Commun. Technol 4:93–97

Shina S, Eksioglu B (2015) An empirical study of RFID productivity in the U.S. retail supply chain. Int J Prod Econo 163:89–96

Tan TH, Chang CS (2010) Development and evaluation of an RFID-based e-restaurant system for customer-centric service. Expert Syst Appl 37(9):6482–6492

Thames L, Schaefer D (2016) Software-defined cloud manufacturing for industry 4.0. Procedia CIRP, 52:12–17

Ustundag A, Tanyas M (2009) The impacts of radio frequency identification (RFID) technology on supply chain costs. Trans Res Part E: Log Trans Rev 45(1): 29–38

Velandia DMS, Kaur N, Whittow WG, Conway PP, West AA (2016) Towards industrial internet of things: crankshaft monitoring, traceability and tracking using RFID. Robot Comput-Integ Manufact 41: 66–77

Wang B, ZHAO JY, WAN ZG, Hong LI, Jian MA (2016) Lean intelligent production system and value stream practice. Trans Econom Manag (ICEM 2016)

Wang YY, Wu YH, Liu YY, Tang AJ (2007) The application of radio frequency identification technology on tires tracking. In: Proceedings of the IEEE international conference on automation and logistics, vol. 1–6, 2927–2930

Wang J, Luo Z, Wong EC (2010) RFID-enabled tracking in flexible assembly line. Int J Adv Manuf Technol 46(1–4):351–360

Zadeh AH, Sharda R, Kasiri N (2016) Inventory record inaccuracy due to theft in production-inventory systems. Int J Adv Manuf Technol 83(1–4):623–631

Zhu X, Mukhopadhyay SK, Kurata H (2012) A review of RFID technology and its managerial applications in different industries. J Eng Tech Manage 29:152–167

Chapter 16
Overview of Cyber Security in the Industry 4.0 Era

Beyzanur Cayir Ervural and Bilal Ervural

Abstract The global development industry is in the midst of a transformation to meet today's more complex and highly competitive industry demands. With the rapid advances in technology, a new phenomenon has emerged in the current era, Industry 4.0. The integration of information technology and operational technology brings newer challenges, especially cyber security. In this chapter, one of the most popular topics of recent times, cyber security issue, has been investigated. The occurrence of the Internet of Things (IoT), has also dramatically altered the appearance of cyber threat. Security threats and vulnerabilities of IoT, industrial challenges, main reasons of cyber-attacks, cyber security requirement and some cyber security measures/methods are discussed with a global perspective involving both the public and private sector in the IoT context.

16.1 Introduction

Industrial revolutions are the most important milestones that have changed the course of human history. According to many researchers, the industrial revolution affects people's lifestyle even more than the science revolutions (Wendt and Renn 2012).

After the discovery of steam power, revolutions have evolved with a rapid change parallel to the needs of each era. Other subsequent revolutions have emerged as electric energy-driven mass production in the early part of the twentieth century, and the use of highly efficient electronic automation in the industrial environment in the 1970s, and finally recently emerged as, Industry 4.0, which has

B.C. Ervural (✉) · B. Ervural
Department of Industrial Engineering, Faculty of Management, Istanbul Technical University, 34367, Maçka, Istanbul, Turkey
e-mail: cayirb@itu.edu.tr

B. Ervural
e-mail: bervural@itu.edu.tr

© Springer International Publishing Switzerland 2018
A. Ustundag and E. Cevikcan, *Industry 4.0: Managing The Digital Transformation*,
Springer Series in Advanced Manufacturing, https://doi.org/10.1007/978-3-319-57870-5_16

come with data-driven production systems, more specifically cyber-physical systems or Internet of Things (IoT) (Rüßmann et al. 2015).

In the present era, the world is at the beginning of the fourth industrial revolution based on IoT. The IoT relies on a variety of enabling technologies such as wireless sensor networks (WSNs), machine-to-machine (M2M) systems, big data, cloud services and smart applications as well as radio frequency identification (RFID) systems (Yang et al. 2010).

The novel change in industries, also known as Industry 4.0, take a great interest from manufacturing companies. It is crucial for manufacturing companies around the world with operational efficiency, productivity and customization features. Industry 4.0 provides dealing with huge data volumes, developing human-machine interactive systems and improving communication between the digital and physical environments (Frost and Sullivan 2017).

Industry 4.0 includes three essential stages: Firstly, getting digital records through sensors that attached to industrial assets, which collect data by closely imitating human feelings and thoughts. This technology is known as sensor fusion. Secondly, analyzing and visualizing step includes an implementation of the analytical abilities on the captured data with sensors. From signal processing to optimization, visualization, cognitive and high-performance computation, many different operations are performed with background operations. The serving system is supported by an industrial cloud to help to manage the immense volume of data. Thirdly, the stage of translating insights to action involves converting the aggregated data into meaningful outputs, such as additive manufacturing, autonomous robots and digital design and simulation. In an industrial cloud, raw data is processed with data analytics applications and then turns into practically usable knowledge.

With Industry 4.0, the interconnected era also dawns. Interconnection provides a link between partners, customers, employees, and systems to accelerate business performance and create new opportunities with collaborating on a shared platform. Interconnection is a requirement for instant access to interdependent and real-time data between industries or between different geographies. The industrial cloud provides a common platform to store data and to collaborate users from various geographies.

Increasing data density with Industry 4.0 and the fusion of information technology and operational technology brings with it newer challenges, particularly of cyber security (Frost and Sullivan 2017). Cyber security is the core issue that all governments follow at the highest level of importance. It is a protection of business information and precious knowledge about a subject or system in digital shape against abuse, unauthorized access and thefts (Kaplan et al. 2011). With expanding network connections, cyberattacks have become more prevalent due to the rising prone to misuse data for different purposes such as financial and strategic reasons.

The boom of new technologies and the increasing societal dependence on globally interconnected technology, the automation, and commodification of the tools of cyberattack, the sophisticated hacker attacks and the low safety measures in the cyber market, are undoubtedly important (Weber and Studer 2016). With the number of potential attackers and the growing size of the network, the tools that potential attackers can use are becoming more sophisticated, efficient and effective. Therefore, it needs to be protected against threats and vulnerabilities in order to achieve the highest potential of IoT (Kizza 2009; Taneja 2013; Abomhara and Kien 2015).

The widespread use of connected devices and services at the IoT has brought about new forms of cyber defense in order to ensure robust security (Sathish Kumar and R. Patel 2014; Abomhara and Kien 2015). Cyber-attacks and threats have increased tremendously in the last decades. Any stakeholder that uses IoT systems is directly or indirectly affected by this matter. Especially large companies are exposed to malicious attacks that result in serious financial burdens in addition to immeasurable losses such as data corruption, system crashes, privacy breaches, prestige, customer, reliability and market losses.

In many organizations, cyber security is primarily considered as a technology issue. Public and private company executives/authorities are aware of the danger, and do not want to allow attackers to access critical business information and employee and customer-specific information (Kaplan et al. 2011). Generally, organizations not officially report the cyber-attacks they are subjected to. To be honest, businesses do not tend to disclose security vulnerabilities that they have to pay ransom for cybercriminals (Kaplan et al. 2011). Most large companies have considerably strengthened their cyber security capabilities in recent years. Millions of dollars have been spent to develop new strategies with technological investments in information technology (IT) security to reduce the risk of cyber-attack.

Internet-based systems will become more attractive for cyber-attacks if IoT achieves a large growth by 2020 (Capgemini 2015). Several companies and organizations have predicted the number of things to connect internets in the coming years (Weber and Studer 2016). According to Gartner predictions, the number of network-connected devices is estimated at 20.8 billion, with Cisco estimated to have about 50 billion IoT connections by 2020, and finally, Huawei's projection shows that by 2025, the number of connections will reach 100 billion. Despite the differences in estimates, the most important result is to expect an important growth. The most obvious inference is that there will be a massive amount of Internet-enabled devices that require a comprehensive protection system soon (Weber and Studer 2016). Figure 16.1 shows the number of connected devices worldwide from 2012 to 2025 (Columbus 2016).

Increasing cyber-attacks in recent years, with victims ranging from individuals to governments around the world, continue to alarm. The year 2014 was declared the Year of the Breach and 2015 was renamed by some industry commentators as the Year of the Breach 2.0. (Weber and Studer 2016). As seen from the general perspective, it is obvious that the cyber-attacks have caused a great deal of damage to the whole world. To prevent cyber-attacks, organizations should educate consumers about the appropriate safety procedures that should be followed while using an IoT system (Capgemini 2015).

This study aims to discuss some issues, such as the increase in data intensity and seriously growing cyber threats due to employing information technology, in the recently emerged concept, Industry 4.0. The widespread use of IoT leads to an

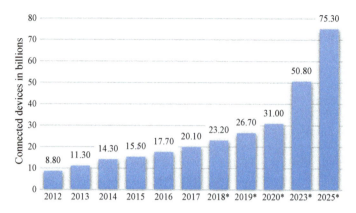

Fig. 16.1 The number of connected devices (Internet of Things) worldwide from 2012 to 2025

increase in the number of interconnected companies, which automatically results in a serious amount of cyber-attacks.

In this chapter, security threats and vulnerabilities of IoT are discussed in Sect. 16.2. Section 16.3 provides industrial challenges. Evolution of cyber-attacks is given in Sect. 16.4. Some cases focus on cyber-attacks and solution approaches are discussed in Sect. 16.5. Strategic principles of cyber security and cyber security measures/methods are respectively provided in Sects. 16.6 and 16.7. The conclusions are presented in Sect. 16.8.

16.2   Security Threats and Vulnerabilities of IoT

There is no single universal consensus on architecture for IoT. Different architectures have been proposed by different researchers (Sethi and Sarangi 2017). In general, the IoT can be divided into four main levels. Figure 16.2 shows both the level architecture of the IoT and some basic components in each level.

– *Perception (Sensing) layer*: The perception layer is also called as 'Sensing Layer'. It composed of physical objects and the sensing devices such as various forms of sensory technologies, RFID sensors. These technologies allow devices to sense other objects.
– *Network layer*: Network layer is the infrastructure to support wireless or wired connections between sensor devices and the information processing system.
– *Service layer*: This layer is to ensure and manage services required by users or applications. It is responsible for the service management and has a link to the database.
– *Application (Interface) layer*: Application or interface layer composed of interaction methods with users or applications. It is responsible for delivering application services to the user.

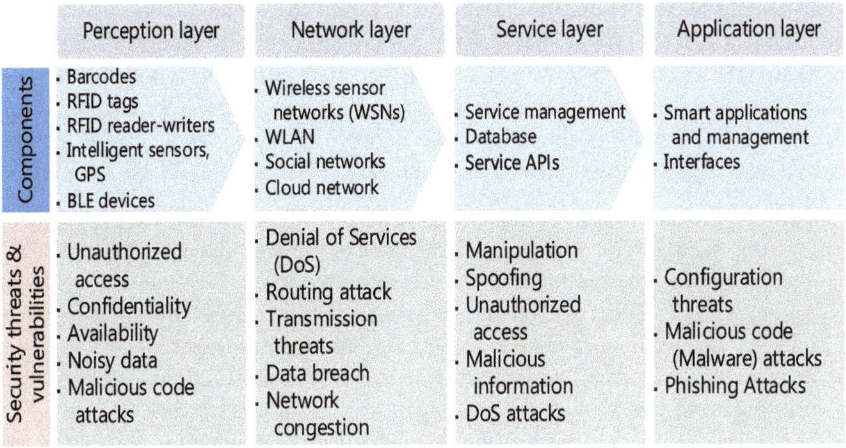

Perception layer	Network layer	Service layer	Application layer
Components			
• Barcodes • RFID tags • RFID reader-writers • Intelligent sensors, GPS • BLE devices	• Wireless sensor networks (WSNs) • WLAN • Social networks • Cloud network	• Service management • Database • Service APIs	• Smart applications and management • Interfaces
Security threats & vulnerabilities			
• Unauthorized access • Confidentiality • Availability • Noisy data • Malicious code attacks	• Denial of Services (DoS) • Routing attack • Transmission threats • Data breach • Network congestion	• Manipulation • Spoofing • Unauthorized access • Malicious information • DoS attacks	• Configuration threats • Malicious code (Malware) attacks • Phishing Attacks

Fig. 16.2 Security threats and vulnerabilities by level

The spread of connected devices in the IoT over a vast area has created a great demand for robust security in addition to the growing demand of millions of connected devices and services worldwide (Abomhara and Kien 2015). The number and complexity of threats and attacks are increasing day by day and tools available to potential attackers are also becoming more sophisticated and effective. Therefore, in order for IoT to reach its full potential, it needs to be strictly protected against threats and vulnerabilities (Kizza 2009; Abomhara and Kien 2015).

The security threats on each layer are different due to its features. Security threats and vulnerabilities according to layers are presented below.

At the perception layer, the intelligent sensors and RFID tags automatically identify the environment and exchange data among devices. The security concerns are an important issue in perception layer. In the perception layer, the majority of the threats comes from external entities, mostly from sensors and other data collection devices. Most of these devices are commonly small in size, inexpensive, and unprotected for physical security (Suo et al. 2012; Kumar et al. 2016; Li 2017). Common threats and vulnerabilities in the perception layer can be summarized as follows:

- *Unauthorized access*: At first node, unauthorized accesses are important threats due to physical capture or logic attack.
- *Confidentiality*: Attackers can place malicious sensors or devices in order to acquire information from the system.
- *Availability*: The system component stops working because it is physically captured or logically attacked.
- *Noisy data (transmission threats)*: The data may contain incomplete information or incorrect information due to transmission over networks covering large distances.
- *Malicious code attacks*: Attackers can cause software failure through malicious code such as virus, Trojan, and junk message.

The network layer connects all things in IoT and allows them to be aware of their environment (Li 2017). The network layer is quite sensitive to attacks because of a large amount of data that it carries. The IoT connects different types of networks, which can cause network security difficulties. Therefore security protection at this level is very important to the IoT. At the network layer, common security threats and vulnerabilities are as follows (Ali et al. 2016; Kumar et al. 2016; Li 2017).

- *Denial of Services (DoS) attack*: Attackers continually bombard a targeted network with failure messages, fake requests, and/or other commands. DoS attacks are the most common threat to the network.
- *Routing attack*: These are attacks on a routing path such as altering the routing information, creating routing loops or sending error messages.
- *Transmission threats*: These are threats in transmission such as blocking, data manipulation, interrupting.
- *Data breach*: A data breach is the intentional or unintentional release of secure or confidential information to an untrusted environment.
- *Network congestion*: A large number of sensor data along with a large number of device authentication can cause network congestion.

In IoT, the service layer relies on middleware technology, which enables communication and management of data in applications and services. Service layer supports and contains the services using application programming interfaces (APIs). In this layer, the data security is crucial and more complicated in comparison to other layers (Li 2017). Some of the common security threats and vulnerabilities in service layer are:

- *Manipulation*: The information in services is manipulated by the attacker.
- *Spoofing*: The information is returned by an attacker to spoof the receiver.
- *Unauthorized access*: Abuse of services accessed by unauthorized users.
- *Malicious information*: Privacy and data security are threatened with malicious tracking.
- *DoS attacks*: A useful service resource is made unavailable by being exposed to traffic above its capacity.

The uppermost layer is the application layer that is visible to the end user. The application layer includes a variety of interfaces and applications, from simple to advanced. The security requirements in the application layer highly depend on the applications. The security threats and vulnerabilities in the application layer are summarized below (Ali et al. 2016; Kumar et al. 2016; Li 2017).

- *Configuration threats*: Failing configurations at interfaces and/or incorrect misconfiguration at remote nodes are the most important threats for this layer.
- *Malicious code (Malware) attacks*: These attacks are intentionally made directly to the software system in order to intentionally cause harm or subvert the intended function of the system.
- *Phishing Attacks*: In the interface layer, attackers may attempt to obtain sensitive information such as usernames, passwords, and credit card details.

The security requirements at all layers are confidentiality, integrity, availability, authentication, non-repudiation and privacy. These requirements are detailed in Sect. 16.6.

16.3   Industrial Challenges

Along with recent developments in IoT platforms, it is almost impossible for the industry to envisage the numerous IoT implementations, given the innovations in the technology, services and continuous needs in the industry (Tweneboah-Koduah et al. 2017). The current application areas include smart manufacturing, smart homes, and smart cities, transportation and warehousing, healthcare, retail and logistics, environmental monitoring, smart finance, and insurance. Investments in IoT solutions by Industry are shown in the Fig. 16.3 (BI Intelligence 2015). Accordingly, the manufacturing sector has an investment volume over 60 billion dollars. Transport and warehousing and information systems are the most invested sectors after the manufacturing sector (Fig. 16.3).

There are security challenges associated with all these application areas. Some of them are very obvious, for example, misuse of personal information, financial abuse. On the other hand, others are more specific depending on the structure of the industry.

With more and more enterprise connected devices being incorporated into the banking sector, the finance industry is faced with an increasing number of ever-evolving cyber security challenges (Craig 2016). Issues of highest concern in financial services industry include protecting privacy and data security, managing third-party risks and stifling compliance regulations.

Although the cyber-attacks have become widespread in the manufacturing industry, recent reports show that energy companies are more prone to these threats, which have become more advanced over the years. At least 75% of companies in

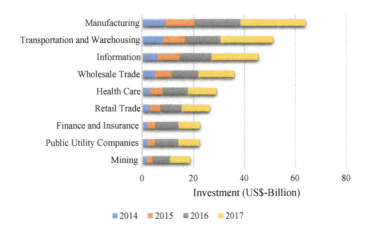

Fig. 16.3 Investments in IoT solutions by industry

the oil, gas and power sectors had one or more successful attacks in 2016. In total, more than 15% of the cyber-attacks are direct attacks on the energy sector (Frost and Sullivan 2017). Challenges of utmost concern in energy industry include protecting privacy and data security, lack of skills and awareness, the integrity of components used in the energy system and increasing interdependence among market players.

The use of IoT in healthcare applications is growing at a fast pace. Many applications such as heart rate monitor, blood pressure monitor and endoscopic capsule are currently in use (Al Ameen et al. 2012). Information security and privacy are becoming increasingly important in the healthcare sector. The storing of digital patient records, increased regulation such as Health Insurance Portability and Accountability Act (HIPAA), provider consolidation, and the increasing requirement for information between patients, providers, and payers point to the need for better information security (Appari and Johnson 2010).

In the transportation industry, rapid developments in technology and widening the connectivity of systems, networks, and devices across transport and logistics

Table 16.1 Challenges according to the industry

Finance	Protecting privacy and data security
	Managing third-party risk: Outsourcing contracts, such as cloud service agreements, impose complex data sharing regulations and generate a host of new cybersecurity challenges
	Emerging and advanced cyber threats
	Regulatory compliance
Energy	Protecting privacy and data security
	Lack of skills and awareness
	Information sharing: Many organizations do not share information about threats or cooperate externally
	Integrity of components used in energy systems
	Increased interdependence among market players
	Alignment of cyber security activities: All activities be aligned and fully integrated with national cyber security
Healthcare	Protecting privacy and data security: Healthcare organizations are required to comply with the Health Insurance Portability and Accountability Act (HIPAA), which requires healthcare vendors to ensure that the privacy of user data is not compromised in any case (Zhang and Liu 2010)
	Medical equipment issues: Healthcare organizations have specialized medical equipment that could pose particular security challenges (Korolov 2015)
	Managing third-party risk: Healthcare organizations are hesitant to move to cloud data protection to ensure that sensitive information is protected without leaving the company network (Zhang and Liu 2010)
Transportation	Protecting privacy and data security especially in the cargo industry (Xu et al. 2014)
	Emerging and advanced cyber threats (DoS attacks, Spoofing attacks) (Warren and Hutchinson 2000)

bring more opportunities in terms of cost, speed, and efficiency. As more devices and control processes are connected on internet environment, more vulnerabilities will emerge. Developing measures against these threats is at the top of the vital issues for the transport sector. Among the major problems in the transportation industry are data security and privacy and emerging and advanced cyber threats.

Some of the industry challenges facing cybersecurity experts are outlined Table 16.1 according to the industries.

16.4   Evolution of Cyber Attacks

The cyber landscape is constantly altering and evolving due to the speed of technological change, the complexity of the attackers, the value of potential targets and the effects of attacks (Weber and Studer 2016).

With the widespread use of computer networks, hackers have taken advantage of network-based services to gain personal benefit and reputation. In a threat environment where security products need to be constantly refined or updated to identify the recent exploitation, the challenge is to find a solution that provides a future-proof defense to ensure lasting network safeguard (Chemringts 2014).

Each organization has digital knowledge and many businesses maintain business transactions and trades with online systems. Most enterprises are open to cyber threats attacking from external and internal boundaries and so, your critical infrastructure needs to be protected (Sheikh 2014).

Cyber security was initially seen as a problem for the IT team, but these days it is an agenda for the entire senior executives. Cybercrime is triggered by sophisticated technologies, the use of mobility, social media, and relatively new trends in rapidly expanding connectivity—all in the hands of organized criminal networks. Under this circumstances, a smart, dynamic and evolutionary approach to cyber security is crucial to stay ahead of cybercrime and competition. Cyber security efforts require protection against a broader range of challenges. It is getting harder with new technologies, trends in mobile usage, social media, well-financed and organized enemies and 24-h attacks. Cyber risks can have a direct impact on everything from stock exchange price to brand reputation, with their more complicated structures (Deloitte 2013).

Figure 16.4 shows how cyber-attacks have evolved over the years and what industry will see in the coming years (Frost and Sullivan 2017).

At the beginning of the 1980s, general cyber-attacks began with password cracking and password guessing methods. Today, directed cyber-attacks occurs with packet spooling, advance scanning, keylogger and denial of service. In future, strategic cyber-attacks are expected to damage strategic points with bots, morphing and malicious codes. Over time, the nature of cyber-attacks has been complicated and extremely sophisticated.

Fig. 16.4 Evolution of
cyber-attacks

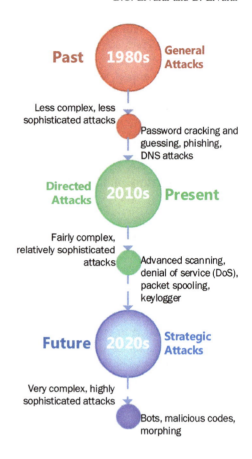

16.5 Cases (Cyber-Attacks and Solutions)

The cyber space is a growing community where everyone can reach each other
independently of time and distance (NATO Review 2013). For this reason, some
people use the cyber space for their own suspicious plans for individuals, corporations, banks, even military and government agencies. In this section, we will
present some important cyber-attacks, which are large-scale cyber terrorism
affecting large masses (Fig. 16.5).

– *Flame*: Flame, also known as Skywiper and Flamer, is a modular computer
  malware discovered in 2012 as a virus that attacks Microsoft Windows operating system computers in the Middle East. When used by spies for espionage, it
  infected other systems via a local area network (LAN) or USB stick with over
  thousands of machines attached to others, educational institutions and government agencies. Skype conversations, keyboard activity, screenshots, and network traffic were recorded. On May 28, 2012, the virus was discovered by

Iranian National Computer Emergency Response Team (CERT), CrySys Lab and the MAHER Center of Kaspersky Lab.

– **July 2009 Cyber Attacks**: A group of cyber-attacks took on major governments' financial websites and news agencies, both United States and South Korea, with releasing of botnet. This included captured computers that lead servers to be overloaded due to the flooding of traffic called DDoS attacks. More than 300,000 computers are hijacked from different sources.

– **The Spamhaus Project**: Spamhaus, considered the biggest cyber-attack in history, is a filtering service used to extract spam e-mails. Thousands of Britons sent Spamhaus every day to determine whether they would accept incoming mail. Spamhaus added Cyberbunker to its blacklisted sites on March 18, 2013; Cyberbunker and other hosting companies have been tasked with recruiting home and broadband routers to hire hackers to abuse botnets to shut down the Spamhaus system.

– **Maroochy Shire sewage spill in Australia (March 2000, Australia)**: The attacker changed the electronic data using the stolen wireless radio, the SCADA controller, and the control software, and all operations failed. It leaded to release up to one million litres of sewage into the river and coastal waters of Maroochydore in Queensland, Australia (RISI 2015).

– **Cyber-attack on Davis-Besse power station of first energy (January 2003, The United States)**: A Slammer worm, entered a private computer network at Davis-Besse nuclear power plant in Ohio and disabled a security monitoring system for about five hours.

– **Public tram system hacked remotely (January 2008, Poland)**: The signaling system on Lodz's tram network was manipulated by a remote control system which was designed by a 14-year old boy utilizing a TV remote control. It

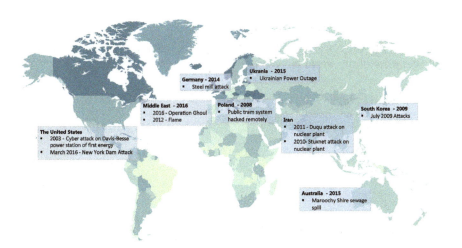

Fig. 16.5 Major industrial cyber-attacks by territories

caused the derailment of four trams and more than a dozen of passengers were injured.

- **Stuxnet attack on Iranian nuclear plant (December 2010, Iran)**: Natanz nuclear plant in Iran was infected by Stuxnet in June 2010, this cyber worm was thought to be a joint effort of Israel and the US but no one took the responsibility of the attack. The worm destroyed 1000 nuclear centrifuges in Tehran and deeply affected the progress of the country because it went beyond just a power plant attack and infected 60,000 computers as well.
- **Duqu attack in Iranian nuclear plant (November 2011, Iran)**: Duqu trojan hits Iran's computer systems. Experts say in a statement to Reuters that Duqu based on Stuxnet is designed to collect data that will facilitate the launch of future cyber-attacks. Stuxnet is intended to disable industrial control systems and may have destroyed some of the centrifuges Iran uses to enrich uranium.
- **Steel mill attack (December 2014, Germany)**: The hackers attacked a steel mill in Germany. By manipulating or disrupting the control systems, it caused major damages in the foundry. Sophisticated attackers entered the steel factory's office network using spear-phishing and social engineering. The production network was reached from this network. With the actions of the attackers, control components and all production machines were cut off.

As can be seen in Fig. 16.6, the cyber-attacks on the Industrial and Commercial IT networks have shown a significant increase in both frequency and intensity over the last four years (Frost and Sullivan 2017).

Attacks targeting industrial control systems (ICS) increased 110% in November 2016 compared to last year, according to IBM management security services data. In particular, the increase in ICS traffic was related to SCADA brute force attacks using automation to guess default or weak passwords. Then attackers can remotely manipulate attached SCADA devices. The United States is the biggest target of ICS-based attacks in 2016 because this attack now has a greater ICS presence than any other country. The top 5 source and destination territories are illustrated in Figs. 16.7 and 16.8, respectively (McMillen 2016).

In the following, several important recent cyber attack cases occurred in the different parts of the world are given.

Fig. 16.6 Industrial IoT system attacks based on years

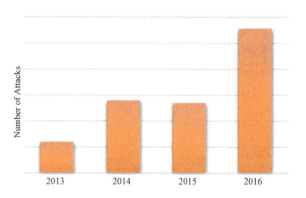

Fig. 16.7 Top 5 source countries

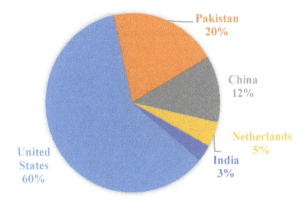

Fig. 16.8 Top 5 destination countries

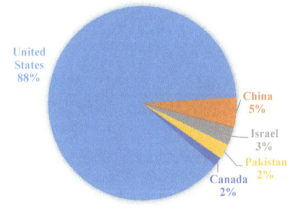

– ***Operation Ghoul***: SFG malware, discovered in a European energy company network in June 2016, has created a back door for targeted industrial control systems. According to security researchers at SentinelOne Labs, the aim is to extract data from the energy network or shut down the energy network. Windows-based SFG malware is created to overcome traditional antivirus software and firewalls.
– ***New York Dam Attack***: In March 2016, computer-based control of a dam in New York was hacked by attackers using cellular modems.
– ***Ukrainian Power Outage***: In December 2015, a power company located in western Ukraine suffered a power outage that impacted a large area that included the regional capital of Ivano-Frankivsk. Three separate energy companies, known as "Oblenergos", were attacked and blocked the power of 225,000 customers. The attack was carried out by hackers using BlackEnergy malware that exploited the macros in Microsoft Excel document. The bug was planted into company's network using spam emails.

The attacks on industrial systems will continue owing to the automation and internet connection increases. This means that the number of such devastating cyber attacks continue to rise and therefore all the damaged organizations will pay a heavy price for the attacks.

16.6 Strategic Principles of Cyber Security

The primary security principles of an efficient IoT security are addressed from six aspects. These principles must be assured for security to be guaranteed in the entire IoT system.

- **Confidentiality**: Confidentiality is the ability to hide information from people who are unauthorized to access it and thus needs protection from unauthorized access (Rodosek and Golling 2013). Confidentiality is an important security feature in IoT. In most situations and scenarios sensitive data for instance patient data, private trade data, and/or military data as well as security credentials and secret keys, must be hidden from unauthorized accesses (Abomhara and Kien 2015).
- **Integrity**: Integrity of information refers to protecting information from unauthorized, unanticipated or unintentional modification. Integrity is a mandatory security property in most cases in order to provide reliable services to IoT users. Different systems in IoT have diverse integrity needs (Abomhara and Kien 2015).
- **Availability**: Availability is the access to information whenever needed by a user of a device (or the device itself). Therefore, the IoT resources must be available on a timely basis to meet needs or to avoid significant losses.
- **Authenticity**: The authenticity property allows only authorized entities to perform certain operations in the network. Different authentication needs require different solutions. Some solutions must be strong control, for example, authentication of finance systems. On the other hand, most must be international, for example, ePassport, while others have to be local (Schneier 2011).
- **Nonrepudiation**: IoT service must provide a trusted audit trail. The property of nonrepudiation presents certain evidence in cases where the user or device cannot deny an action, for instance, payment action.
- **Privacy**: Privacy is an entity's right to determine the degree to which it will interact with its environment and to what extent the entity is willing to share personal information with others (Abomhara and Kien 2015).

16.7 Cyber Security Measures

Cyber security measures must be taken in the future to reduce cyber risks. We will explain some basic cyber security precautions/measures as much as possible to prevent all possible attacks.

- Do not allow to connect directly to a machine on the control network, on a business network or on the Internet. Organizations may not realize this connection exist, a cyber attacker can find a gap to access and exploit industrial control systems to give rise to physical damages. For this reason channels between the devices in the control system and other network devices must be removed from the center to reduce network openings (WaterISAC 2015).
- A firewall is a software program or hardware device that filters incoming and outgoing traffic between different parts of a network or between a network and the Internet. Do not allow a threat to easily reach your system by reducing the number of routes in your networks and applying security protocols to the routes. Establishing network boundaries and segments gives an organizational authority to implement both detective and protective controls on the infrastructure. The monitoring, restriction, and management of communication flows provide the practical capability for basic network traffic (especially for traffic that exceeds a network limit) and define abnormal or suspicious communication flows.
- Remote access to a network using some conservative methods like Virtual Private Network (VPN) provides big advantages to the end users. This remote access can be strengthened by reducing the number of Internet Protocol (IP) addresses that can access it by using network devices and/or firewalls to identified IP addresses.
- Role-based access control allows or denies access to network resources based on business functions. This limits the ability to access files or system parts that individual users (or attackers) should not be able to access.
- Applying strong passwords is the easiest way to strengthen your security. Hackers can use software tools that are easily accessible to try millions of character combinations to gain unauthorized access—it is called brute force attack. According to Microsoft, you should definitely avoid using personal data (such as date of birth), backwards-known words, and character or number sequences that are close together on the keyboard (BI Intelligence 2010). Create a password policy to help employees monitor best practices for security. Various technology solutions can be supported to enforce your password policy, such as scheduled password reset (Nibusiness 2017).
- Many Internet-enable devices include hard-coded default credentials. Such identity information is often freely available on the Internet and is widely known by people. Most malware targeting IoT devices is only performed by attackers using default credentials. According to Microsoft, you should definitely avoid using personal data (such as date of birth), backwards-known words, and character or number sequences that are close together on the keyboard.
- It is important to ensure awareness of vulnerability and application of required patches and updates. To protect an organization from opportunistic attacks, a system must be implemented to monitor and enforce system settings and updates. Organizations should consider updating system and software settings automatically to avoid missing critical updates.
- Your employees are responsible for helping to ensure the safety of your business. It is very important to give your employees information about safe online

habits and proactive defense and give them regular cyber security awareness and training.

- Due to the portable nature, there is a greater risk of laptop computers. It is important that you take extra steps to protect sensitive data. It is important that you take additional steps to protect sensitive data. Encrypting your laptop is the easiest way to take precautions. Encryption software changes the way information appears on the hard drive, so it cannot be read without the correct password (BI Intelligence 2010).
- Nowadays smartphones are in the center of everything, so it should be considered that they are valuable as much as company computers in case of lost or stolen. Encryption software, password protection, and application of remote wiping are very effective securing methods for smartphones to all possible attacks (BI Intelligence 2010).
- Organizational leaders generally do not know the threats and needs of cyber security. Incorporating managers into the scope of cyber security helps corporations with cyber security issues in interactions with external stakeholders (WaterISAC 2015).
- Nevertheless, administrators should not rely solely on anti-virus software to detect infections. Firewalls, intrusion detection and prevention sensors and logs from the servers should be monitored in terms of infection indication. Incident response plans are a critical but not yet sufficiently used component of emergency preparedness and flexibility. An effective cyber security measure will limit the damage, increase the trust of partners and customers, and reduce recovery costs and time (WaterISAC 2015).

16.8   Conclusion

The development of new digital industrial technology led to the emergence of Industry 4.0, the fourth wave of the industrial revolution. Industry 4.0 deals with huge data volumes, developing human-machine interactive systems and improving communication between the digital and physical environments, namely in the IoT context.

With Industry 4.0, the combination of information technology and operational technology have brought new challenges. Cyber security is the main issue that all governments in the world have made a great deal of effort against cyber security attacks. By 2020, more than 50 billion IoT devices have revealed that how important cyber security is.

In this chapter, the concept of cyber security is investigated from a comprehensive perspective, based on the context of IoT, involving many stakeholders from different sectors of the global world. The requirement of cyber security, security threats, and vulnerabilities of IoT, the evolution of cyber-attacks and cyber security measures are discussed and supported with some graphs, figures, tables and studies in the literature.

As new platforms and operating systems for connected devices continue to evolve, security budgets are expected to grow exponentially for all organizations. The future of the cyber security strongly depends on considering threat landscapes and emerging trends in technology related to big data, cognitive computing, and IoT.

References

Abomhara M, Kien GM (2015) Cyber security and the internet of things: vulnerabilities, threats, intruders and attacks. J Cyber Secur 4:65–88

Al Ameen M, Liu J, Kwak K (2012) Security and privacy issues in wireless sensor networks for healthcare applications. J Med Syst 36:93–101. doi:10.1007/s10916-010-9449-4

Ali I, Sabir S, Ullah Z (2016) Internet of things security, device authentication and access control: a review. Int J Comput Sci Inf Secur 14:456

Appari A, Johnson ME (2010) Information security and privacy in healthcare: current state of research. Int J Internet Enterp Manag 6:279. doi:10.1504/IJIEM.2010.035624

BI Intelligence (2010) 10 data-security measures. Bus. Insid. Digit

BI Intelligence (2015) The enterprise internet of things market—business insider. http://www.businessinsider.com/the-enterprise-internet-of-things-market-2015-7. Accessed 19 Jun 2017

Capgemini (2015) Securing the internet of things opportunity: putting cybersecurity at the heart of the IoT | Capgemini Worldwide

Chemringts (2014) The evolution of cyber threat and defence. UK

Columbus L (2016) Roundup of internet of things forecasts and market estimates In: Forbes. https://www.forbes.com/sites/louiscolumbus/2016/11/27/roundup-of-internet-of-things-forecasts-and-market-estimates-2016/#7eba4b9a292d. Accessed 20 Jun 2017

Craig D (2016) Five cybersecurity challenges facing financial services organizations today. IBM Secur Intell

Da Xu L, He W, Li S (2014) Internet of things in industries: a survey. IEEE Trans Ind Inform 10:2233–2243. doi:10.1109/TII.2014.2300753

Deloitte (2013) Risk angles—five questions on the evolution of cyber security. United Kingdom

Frost and Sullivan (2017) Cyber Security in the Era of Industrial IoT. Frost & Sullivan White Paper, Germany

Kaplan J, Weinberg A, Sharma S (2011) Meeting the cybersecurity challenge. Digit. McKinsey

Kizza JM (2009) Guide to computer network security. Springer, Berlin

Korolov M (2015) Healthcare organizations face unique security challenges | CSO Online. CSO

Kumar SA, Vealey T, Srivastava H (2016) Security in internet of things: challenges, solutions and future directions. In: 2016 49th Hawaii international conference on system sciences (HICSS). IEEE, pp 5772–5781

Li S (2017) Security requirements in IoT architecture. In: Securing the internet of things, pp 97–108

McMillen D (2016) Attacks targeting industrial control systems (ICS) up 110 percent. IBM

NATO Review (2013) The history of cyber attacks—a timeline. http://www.nato.int/docu/review/2013/Cyber/timeline/EN/index.htm. Accessed 20 Jun 2017

Nibusiness (2017) Common cyber security measures. In: nibusinessinfo.co.uk. https://www.nibusinessinfo.co.uk/content/common-cyber-security-measures. Accessed 20 Jun 2017

RISI (2015) RISI—the repository of industrial security incidents. http://www.risidata.com/Database/Detail/maroochy-shire-sewage-spill. Accessed 20 Jun 2017

Rodosek GD, Golling M (2013) Cyber security: challenges and application areas. Springer, Berlin, pp 179–197

Rüßmann M, Lorenz M, Gerbert P et al (2015) Industry 4.0: the future of productivity and growth in manufacturing industries

Sathish Kumar J, Patel DR (2014) A survey on internet of things: security and privacy issues. Int J Comput Appl 90:20–26. doi:10.5120/15764-4454

Schneier B (2011) Secrets and lies : digital security in a networked world. Wiley, Hoboken

Sethi P, Sarangi SR (2017) Internet of things: architectures, protocols, and applications. J Electr Comput Eng 2017:1–25. doi:10.1155/2017/9324035

Sheikh S (2014) Evolving cyber security—a wake up call… In: Marsh National Oil Conference. Dubai

Suo H, Wan J, Zou C, Liu J (2012) Security in the internet of things: a review. In: 2012 international conference on computer science and electronics engineering. IEEE, pp 648–651

Taneja M (2013) An analytics framework to detect compromised IoT devices using mobility behavior. In: 2013 international conference on ICT convergence (ICTC). IEEE, pp 38–43

Tweneboah-Koduah S, Skouby KE, Tadayoni R (2017) Cyber security threats to IoT applications and service domains. Wirel Pers Commun 1–17. doi:10.1007/s11277-017-4434-6

Warren M, Hutchinson W (2000) Cyber attacks against supply chain management systems: a short note. Int J Phys Distrib Logist Manag 30:710–716. doi:10.1108/09600030010346521

WaterISAC (2015) 10 Basic cybersecurity measures—best practices to reduce exploitable weaknesses and attacks

Weber RH, Studer E (2016) Cybersecurity in the internet of things: legal aspects. Comput Law Secur Rev 32:715–728. doi:10.1016/j.clsr.2016.07.002

Wendt H, Renn J (2012) Knowledge and science in current discussions of globalization. In: The globalization of knowledge in history. Edition Open Access

Yang D-L, Liu F, Liang Y-D (2010) A survey of the internet of things. In: Proceedings of the 1st International Conference on E-Business Intelligence (ICEBI2010). Atlantis Press

Zhang R, Liu L (2010) Security models and requirements for healthcare application clouds. In: 2010 IEEE 3rd international conference on cloud computing. IEEE, pp 268–275

Index

A. Ustundag and E. Cevikcan, *Industry 4.0: Managing The Digital Transformation*,
Springer Series in Advanced Manufacturing, https://doi.org/10.1007/978-3-319-57870-5

CPSIA information can be obtained
at www.ICGtesting.com
Printed in the USA
LVHW06*0502170718
583816LV00001BA/93/P

9 783319 578699